The Biology and Management of Mountain Ungulates

THE BIOLOGY AND MANAGEMENT OF MOUNTAIN UNGULATES

Edited by

SANDRO LOVARI

Istituto di Zoologia, Universitá di Parma, 43100 Parma, Italy
and currently at the Sub-department of Animal Behaviour,
University of Cambridge, UK

Published under the auspices of the Caprinae Specialist Group
(Species Survival Commission of the International Union for the
Conservation of Nature, CH – Gland)

CROOM HELM

London • Sydney • Dover, New Hampshire

Croom Helm Ltd, Provident House, Burrell Row,
Beckenham, Kent BR3 1AT
Croom Helm Australia Pty Ltd, First Floor, 139 King Street,
Sydney, NSW 2001, Australia
Croom Helm, 51 Washington Street, Dover, New Hampshire 03820, USA

British Library Cataloguing in Publication Data

The Biology and management of mountain ungulates.
1. Lovari, Sandro
I. Title
599.7 QL737.U4

ISBN 07099-1688-4

Library of Congress Cataloging in Publication Data

Main entry under title:
The Biology and management of mountain ungulates.
Proceedings of the fourth International Conference on Chamois and Other Mountain Ungulates, held in Pescasseroli,—Abruzzo National Park, Italy, June 17-19, 1983.
"Published under the auspices of the Caprinae Specialist Group (Species Survival Commission of the International Union for the Conservation of Nature, CH - Gland)"
Includes index.
1. Mountain goat—Congresses. 2. Wildlife management—Congresses. I. Lovari, Sandro, 1946-
II. International Conference on Chamois and Other Mountain Ungulates (4th : 1983 : Pescasseroli, Italy)
QL737.U53B54 1985 599.73'5 84-21239

ISBN 0-7099-1688-4

Typeset by Columns of Reading
Printed and bound in Great Britain

Contents

Part Two: Ecology and Behaviour

Part Three: Management and Conservation

Part Four: Management of Health and Disease

Poets may praise the deer and nightingale; I celebrate the wild goat.

(G.B. Schaller, 1980, *Stones of Silence*. The Viking Press, New York)

Preface

Mountain ungulates have long been disregarded by scientists as appropriate research animals. It is difficult to say why this has been so, as the highly selective environment in which these animals evolved has produced specialised, yet resilient, species which, by their own biological features, make stimulating subjects of study. Probably the rugged terrain and the cold, wet climate of the mountains have somehow militated against the development of extensive studies such as those conducted on species living in lowlands or plains habitats. It may seem paradoxical, but on the whole we definitely know more about the biology of some exotic African ungulates than about that of most European mountain species!

Although for the local villagers the ibex and the chamois are no strangers, in fact most people only know the ibex through picture postcards and the chamois by the resilient and soft piece of bovine skin used to clean glasses and lenses. For centuries those who knew these species best were probably those who liked to hunt and kill them.

Only in the last few decades, have there been a number of studies of mountain ungulates, both in the Old and the New Worlds. During this time, the quality of research has greatly improved and there is now an increasing number of scientists working in this field.

So far not much has been published in English on European

mountain ungulates. Therefore, this collection of papers and reports is probably the most extensive available in this language and it stems out of the conference — of which this book is the proceedings — held in Pescasseroli, Abruzzo National Park, Italy (the *4th International Conference on Chamois and Other Mountain Ungulates*, 17-19 June 1983). Actually, 86.2 per cent of the papers submitted had to be revised later on, and several invited articles were added. So, in fact, the contents of this volume are updated to 1984. As literature is already available in English on American mountain ungulates, no effort was made to include papers on such species. Rather, emphasis has been put on the European ones; this also keeps the size of the volume within reasonable limits.

Each full paper was screened by at least two referees, whereas most short notes, recognisable because they lack an 'abstract' and major sub-headings, underwent only editorial revision.

Two-thirds of the book is concerned with the biology of mountain ungulates and one-third is directly relevant to management.

At present, the status of wildlife as a whole is poor; most wild animals are continually pushed into smaller and smaller areas by their main enemies: 'development' and 'progress'. Therefore, more than ever before, population management must be based on sound scientific data. This is a relatively new concept and it often heavily interferes with the traditional aspects of hunting. Both the number and some of the contents of the management papers reflect this trend.

Let us hope that the era of sound, objective management (conservation included) has started for mountain ungulates: they badly need it.

Acknowledgements

I wish to thank: Danilo Mainardi, for his encouragement throughout both the organisation of the *4th International Conference on Chamois and Other Mountain Ungulates* and the laborious editing of this volume; Rosario Fico, Stefano Maugeri, Marianna Patalano and in particular Riccardo Nardi and Stefano Tribuzi for their invaluable friendly help in organising the final stages of the conference; all the referees I consulted (their names are listed on page xii), in particular Hartmut Gossow, Kurt Onderscheka and Wolfgang Schröder for their extensive advice on the papers submitted; furthermore, W. Schröder and the researchers of the Abteilung für Wildforschung, Munich University, for their encouragement and help in the initial planning of the conference; the secretaries of the Istituto di Zoologia, Parma University, for re-typing a large part of the manuscripts.

I wish to thank also Mary Horgan for her help in the final editing of this book.

The Abruzzo National Park Agency, the Istituto di Zoologia of Parma University and the Chamois Survival Group — Italy endorsed the conference, whereas the universities of Parma and Bologna provided generous financial support without which this volume could hardly have been produced. The help of Danilo Mainardi, Emilio Baldaccini and Carlo Ferrari proved crucial in this connection and I am deeply grateful to them all.

S.L.

Consultant Editors for Papers in this Volume

(1) F.J. Ayala (Davis, USA)
(2) T. Balbo (Turin, Italy)
(3) C. Berducou (Toulouse, France)
(4) C.M.H. Clarke (Christchurch, New Zealand)
(5) T.H. Clutton-Brock (Cambridge, UK)
(6) E. Duffey (Huntingdon, UK)
(7) V. Geist (Calgary, Canada)
(8) A.W. Gentry (London, UK)
(9) B. Georgii (Oberammergau, FRG)
(10) H. Gossow (Vienna, Austria)
(11) C.P. Groves (Canberra, Australia)
(12) P. Grubb (London, UK)
(13) A. Krämer (Frauenfeld, Switzerland)
(14) B. Kurtén (Helsinki, Finland)
(15) C. Ngampongsai (Bangkok, Thailand)
(16) B. Nievergelt (Zurich, Switzerland)
(17) H. Prosl (Vienna, Austria)
(18) K. Onderscheka (Vienna, Austria)
(19) C.L. Remington (New Haven, USA)
(20) J. Schröder (Munich, FRG)
(21) W. Schröder (Munich, FRG)
(22) H. Wiesner (Munich, FRG)
(23) U. Wotschikowski (Oberammergau, FRG)

PART ONE

Evolution and Systematics

CHAPTER 1

On Evolutionary Patterns in the Caprinae with Comments on the Punctuated Mode of Evolution, Gradualism and a General Model of Mammalian Evolution

V. GEIST

Faculty of Environmental Design, University of Calgary, Calgary, Alberta T2N 1N4 (Canada)

Abstract

In developing evolutionary relationships between species, it is necessary to establish criteria other than those from fossil evidence, particularly where the fossil record of a group is inadequate. A general model of mammalian evolution was developed and is applied to caprids in this paper. It is shown that caprids followed not only the general model in their late Tertiary-Pleistocene radiations, but that one can also use the model to elucidate relationships in such confusing genera as *Capra* and *Ovis*. The general model used is a development from the dispersal theory, a theory explaining speciation in the ice age mammals, which predicts a punctuated-equilibrium mode of speciation, but also a gradual evolutionary change via efficiency selection during the life of a species.

Introduction

The evolution and taxonomy of caprids have been ably discussed by Schaller (1977) and I shall make his thorough review my point of departure. The general picture which Schaller presents is acceptable, but some of the details are contentious. In preparation for re-examining caprid phylogeny I have examined the evolution of other ice age mammals in order to discern the rules by which this evolution proceeds. In particular, I focused on Old World deer whose fossil record is better than that of the caprids, especially since antlers become fossilised and horns do not. One can generate a model of late-Tertiary and Pleistocene evolution of large mammals (see Geist, 1978, 1983) and the caprids, as I shall show, fit this general model wherever it can be tested.

I may be permitted to elucidate some of my concerns and approaches in deciphering evolutionary patterns. Although I do agree that, ideally, the fossil record should be the ultimate means of reconstructing phylogeny, in practice this is not always possible. Continuous advances in our understanding of Pleistocene geology call for a re-examination of older records, while the poor quality of fossil material coupled with ignorance about the biology of the taxa involved can lead to severe misidentification. I am therefore concerned with developing phylogenetic pictures if possible by means other than palaeontology and testing the fossil record against such. For instance, Old World deer show very well in the fossil record an orderly progression in evolution from small-bodied, tusked forms in the mid-Tertiary to large-bodied complexly antlered forms in the Villafranchian and Pleistocene. One can next compare living members of this subfamily against comparable members from the fossil record, and discover that these grades show progressions in display organs other than horn-like organs, as well as clear ecological relationships. One can thus identify primitive and advanced even in the absence of fossils. One can also see a very clear zoogeographic pattern to evolutionary radiation, which was the factor that led me to develop a theory about the manner of speciation which I first called 'glaciation' theory (Geist, 1971a) and then changed to 'dispersal' theory (Geist, 1971b, 1978, 1983). This theory, developed prior to the controversy about 'punctuated equilibrium' versus 'gradualism' in evolution (Gingerich, 1977; Gould & Eldredge, 1978), predicts rapid speciation events, followed not by evolutionary stasis, but by very slow changes mostly in organs of food procurement and processing so that the individuals become increasingly more efficient at reproducing at lower costs. There is therefore some 'gradualism' in evolution, but it is specified as to its nature: it is the evolutionary fine-tuning towards greater efficiency and the ridding of costly structures and behaviours.

The dispersal theory, which underlies the model summarised below, is an arbitrator of evolutionary speculation. Is the argali primitive or advanced over urials? It is *advanced* by at least three criteria:

1. enlarged social organs at head and rear pole with reduced tail;
2. larger body size;
3. living in more extreme habitats at higher altitude and latitude.

What about 'argali type' sheep appearing in Europe and China in early mid-Pleistocene times followed by small, urial-like sheep in the late Pleistocene (Herre & Kesper, 1953; Korobitsyna, Nadler, Vorontsov & Hoffman, 1974)? The inference here that urials arose from argalis (Schaller, 1977:333) is not admissible for several reasons: every major group of glacial advances will generate new sheep, all of which are likely to start out 'big' and therefore appear 'argali like' (large American fossil bighorns have also been postulated to be argalis by Stokes & Condie, 1961). Every major set of glacial events will also cause extinctions. In the last 700,000 years at least nine major glacial pulses have occurred, and since the onset of major glaciations about two million years ago some forty-one (Nilsson, 1983:458). In Europe the record for *Ovis* is highly discontinuous and sporadic; therefore, there is little chance that large mid-Pleistocene sheep, let alone earlier ones, remained *in situ* to evolve into urials. It is more likely that urials have repeatedly given rise to radiations of large sheep which radiated into areas made temporarily into sheep habitat by glacial events, and that these radiations became usually extinct by reversals in glacial events. However, remnants of these radiations could on occasion survive provided they penetrated far enough to find areas which allowed survival even in times of sheep habitual contraction. We expect therefore some pockets of very old *Ovis* close to the fringe of areas greatly affected by glacial events. Also, the dispersal theory predicts that these old sheep will be geographically close to their ancestral form, and that their ancestral form will be in a warmer climate. This form I have identified on morphological criteria as *Ammotragus* (Geist, 1971a), a conclusion supported by the chromosome work of Nadler, Hoffman & Woolf (1974). The most primitive *Ovis* alive are the bibbed, 2n=58 chromosome urials.

Could not evolved argalis invade urial range and outcompete them? No. Primitive forms have specialised in resource competition, and have existed for longer periods in a given area, and therefore have longer histories of efficiency selection. Argalis moving on to urial range would find very little forage for growth and reproduction; I predict that such dispersers will return to where they came from. Also, dispersers, by virtue of greater size and activity, would be particularly vulnerable to food shortages and thus could not dwell in an area avidly exploited by maintenance phenotype sheep. Only in America has a northern fauna invaded southern regions, but only during or after the native Rancho-La Brean fauna collapsed. Ironically, when an old desert genus of America, *Equus* (*Asinus*), in the form of the burro was introduced to ranges occupied by bighorn sheep, it rapidly multiplied, thrived and outcompeted bighorns (Bailey, 1980). In addition, advanced forms, by having evolved in extreme climates are likely to have lost indigenous parasites and the immunity to them. Upon entering the range of a more primitive form in somewhat more benign climates, they may succumb to the diseases and parasites of the old form. The rule I am concerned with here is that an advanced form will skirt around the distribution area of a more primitive form, but will not invade and displace the primitive form. Primitive forms are more likely to survive geoclimatic events than are advanced ones. The dispersal theory predicts, and the punctuated equilibrium model postulates via empirical data, that following speciation a species is more or less frozen in external appearance despite some 'fine-tuning' to make it more efficient. The postulate that argalis gave rise to urials is thus not acceptable. It is quite likely that large argalis could give rise to small argalis, just as big bighorns have given rise to small bighorns in America (Harris & Mundel, 1974), but it is not likely that there would be a *reverse-evolution*.

There are suggestions by palaeontologists about caprid evolution that become untenable. *Tossunnoria* has been suggested as the ancestor of sheep and goat (Thenius & Hofer, 1960), but *Tossunnoria* is even more primitive than the tahr (*Hemitragus*) let alone *Ammotragus*, and is found in China, the wrong geographical location. It is part of the Villafranchian caprid radiation. The tahr is suggested as the ancestor of *Capra* by Schaller (1977), but *Ammotragus* is closer morphologically, behaviourally, in karyotype and by zoogeographical criteria. The large bighorns that initiated colonisation of the southern Rockies about 20,000 years ago (Harris & Mundel, 1974) have been placed into a separate species *Ovis catclawensis* (Stokes &

Condie, 1961; Stock & Stokes, 1969). The dispersal theory predicts that they are only larger phenotypes of the species *Ovis canadensis*; they are only some 12-20 per cent larger in linear dimension (50-70 per cent larger in mass) which is within the range of phenotypic size difference of other ungulates such as red deer (Vogt, 1936; Beninde, 1937; Geist, 1983)

A Model of Pleistocene Radiation in Mammals

Modern lineages of large mammals take their origin in the following.

1. A small-bodied, tropical, resource defender from dense forests or shrub. As a resource defender it has short, often sharp weapons that maximise surface damage to the body and defence via evasion or dermal shield. It is either not gregarious or gregarious in the context of resource defence. There is little differentiation of head and anal pole. Food habits tend to be catholic and the dentition low crowned. Sexual dimorphism is low.
2. The primitive body plan is retained by the tropical radiation, although some departure arises if the forest/steppe ecotone is breached. In the tropical radiation species differentiate by ecological specialisation, resulting in a series of increasing body size in which the largest is an ecologically highly specialised giant, but one incorporating the primitive body plan.
3. If tropical savannah is invaded, great changes follow. Resource defence is abandoned, gregariousness as an anti-predator strategy is assumed, weapons change from highly damaging types to such that do not pierce conspecifics in the normal, everyday social interactions, or which can be used for wrestling. Sexual dimorphism increases, the dentition adapts to grasslike vegetation and becomes selenodont, social organs become conspicuous, horn-like organs enlarge and rear poles begin to differentiate. The signals used in communication become adapted to life in the open.

The tropical radiations thus produce species with a primitive body plan although it may be greatly altered via ecological adaptations. For example: the common giraffe has the same body plan as the okapi, which is in its structure as primitive as the earliest of mid-Tertiary giraffes (Churcher, 1978). Both possess short ossicones, both have massive dermal armour, but the body of the giraffe is altered in the direction of specialising on feeding in trees. By comparison, giraffes with an advanced body plan were the extinct sivatheres, short-necked, palmate-horned, large-bodied grazing giraffes that occupied dry steppes. Primitive is used here in the sense of original; advanced in the sense of later arising or derived.

4. Evolutionary radiation proceeds into landscapes of increasing climatic severity, following first a gradient of increasing aridity and then a gradient of increasingly lower annual mean temperatures, and increasing seasonal temperature differences. This results in:

a. a gradient of species from savannah, through steppe to desert.
b. a gradient of species (or species clusters) moving from tropical to cold climates. The end products of this radiation are the grotesque giants in alpine, arctic and periglacial environments.

5. Regardless of evolutionary grade, colonisation of islands leads to dwarfing. I shall not deal here with this phenomenon.

6. Species fluctuate in body size in time and space. Historically, the largest body sizes are shown by the earliest members of a species; body size fluctuates but tends to decline with the age of a species. Locally body size varies with the productivity pulse of a region (Margalef, 1963), the predation pressure or the expansion of a population. The phenotypically largest body size is achieved during colonisation when material resources are superabundant and not subject to intraspecific (or interspecific) competition (see Geist, 1978, 1983)

7. Historically, there is not only a decline in body size in a species, but also an increase in organs that are associated with food acquisition or processing. Thus cheek teeth may become increasingly molarised, larger and higher crowned. This suggests that there is selection for increasingly more efficient chewing and liberation of energy and nutrients from coarser food. That is, there is selection for efficiency. The progressive shortening of the nasal bones of moose (*Alces*; Kurten & Anderson, 1980) with a concomitant elongation of the premaxillaries indicates an

enlargement of the proboscis-like nasal organ on which moose depend for aquatic feeding and rapid, selective browsing. Another example is the progressive increase in the size of teeth in African pigs. However, in some forms teeth may also decline in relative size (Gingerich, 1977) in response to specialisation to softer forage. In Pleistocene mammals tooth size may also parallel body size changes with stadials and interstadial (Tchernov, 1981).

8. Glaciations and concomitant environmental changes are the driving force of Pleistocene evolution leading to extinction, recolonisation and speciation.

9. Speciation is linked to dispersal into large areas of favourable habitat and the selection pressures associated with colonisation. Food abundance triggers the development of *dispersal phenotypes* with an r-strategy of reproduction. Release from competition for material resources leads to intense social competition among males, selecting for larger bodies, sturdier weapons, better bluffing ability. Social selection is intense, non-Darwinian evolution is high; and there may be selection for environmental factors such as new landforms, predators, climatic events, forage plants, etc. Colonisation is thus a period of maximum genetic change. Strong directional selection can lead to phenodeviants, which greatly increases variability in the speciation stage, and may result in the rearrangement of the chromosomal architecture of the emerging new form. Therefore, one expects great variability of individuals during and following colonisation.

When a vacant habitat is filled, competition for material resources sets in, individuals assume a *maintenance phenotype* and efficiency selection sets in, fine-tuning the characters evolved (see Geist, 1978, 1983).

The process just described will not explain the evolution of tropical ecological specialists which retain a primitive basic body plan. Ecological specialisation would appear to be the result of severe competition for material resources. Such will not permit the phenotypic development of luxury organs, unless there is a periodic relief from such competition, such as a seasonal superabundance of forage during the seasonal productivity pulse. If there is no such annual vacation from competition, body size will reflect strictly the determinants of ecological adaptation.

Even if the act of speciation is initiated via dispersal into a vacant habitat, perpetual competition for resources will quickly reduce the size of social organs such as horns or antlers to a minimum. How a string of species evolves in the tropics which retain basically similar body plan, but differ in ecological specialisation, still needs explaining. We are thus dealing with more than one mode of speciation.

10. Since speciation proceeds together with colonisation, it generates zoogeographical patterns of species distribution. Closely related species will tend to be close together geographically; evolutionary advancement may be traced geographically without any apparent ecological or climatic gradient; groups of unrelated species may show close parallels between evolutionary advancement and dispersal; recent postglacial radiations are more likely to have a contiguous distribution while older ones are likely to be fragmented and reduced.

11. In general, forms evolved in cold and temperate climates have not displaced relatives from warmer climates. Only in North America can we see today a northern fauna in southern latitudes; this is largely the result of megafaunal extinction at the end of the Wisconsin glaciation which permitted the northern fauna to move in and fill a vacant landscape. It is here one can study cold-adapted mammals adapting to southern climates, an exceptional situation for ice age mammals.

How Well Do the Caprids Fit?

The scattered fossil record of the caprids, the sheer evolutionary turmoil of the ice ages, the inherent difficulties of correctly identifying fragmented bones and the great diversity of caprids combine to obscure any palaeontological picture of an orderly progression from the Tertiary to the Recent. From the fossil record alone we cannot assemble a pattern as shown for horses, or giraffes or Old World deer. However, caprids do fall into the expected pattern if we arrange them by body size and climatic severity (Figure 1.1). Serow and goral represent the original stage of resource defenders from warm climates and we expect them to be very old. The goral is found in the Villafranchian, the 'dawn of the ice ages' in Europe (Kurten, 1968). A primitive small-bodied mid-Tertiary ancestor is missing today. We expect it to be

Figure 1.1: An Approximate Alignment of Rupicaprids and Caprids by Body Size and Climatic Extreme. A. goral (*Nemorhaedus*), B. mainland serow (*Capricornis sumatraensis*), C. wild goat (*Capra aegagrus*), D. aoudad or barbary sheep (*Ammotragus lervia*), E. chamois (*Rupicapra rupicapra*), F. markhor (*Capra falconeri*), G. tahr (*Hemitragus jemlahicus*), H. takin (*Budorcas taxicolor*), J. mountain goat (*Oreamnos americanus*), K. Siberian argali (*Ovis ammon ammon*), L. musk ox (*Ovibos moschatus*) (from Geist, 1983)

serow-like. The serow would be the primitive giant, comparable to the sambar of Old World deer, the sloth bear (*Melursus*) among the ursids, the gorilla and orang-utan of the anthropoids, or the gaur (*Bibos gaurus*) among cattle.

The step from life in cover to life in the open and concomitant entry into gregarious existence without resource defence has been taken repeatedly. It is taken by the tahr (*Hemitragus*), and herewith to the evolution of the Caprini proper. We expect the tahr to be old; in Europe it appears in the Villafranchian (Astian) and repeatedly in the major ice ages (Kurten, 1968). We expect it to precede all other genera and, if the Astian record is correct, then it does. The great age of *Hemitragus* is indicated zoogeographically by the far-flung distribution of its existing species, and the intervening barriers. As to the fossil evidence for *Ovis* and *Capra*, it is an uncertain record at best. Both genera became common in the late Pleistocene, but there are sporadic records right back to the Villafranchian. The status of these records needs a re-examination, because in the Villafranchian there was an outburst of rupicaprid speciation, and convergent evolution, alas, is very common in mammals (see Geist, 1978: 172-84).

Other lineages which have taken the step from forest to open landscapes in the Caprinae are the musk oxen, the takin, and probably a number of Villafranchian genera as one can judge from their armaments. Even the chamois, having become a gregarious form, has altered the daggerlike horns of the Rupicaprini so that at least on a simple head-butt they will not pierce an opponent. The sharp horns of the mountain goat have remained as this species relapses into resource defence and territoriality during severe winters (Kuck, 1977).

We expect that species from extreme environments evolved later than did those from warm climates, and in the case of the Rupicaprini this appears to be indeed the case. *Rupicapra* and *Oreamnos*, both periglacial, montane species, probably evolve in the long Riss/Illinoian glaciations as they appear in the deposits of the long interglacial preceding the last major glaciations (Würm/Wisconsin) in Europe and North America (see Kurten, 1968; Kurten & Anderson, 1980).

Musk oxen, however, which are far more specialised to cold climates, appear much earlier than the cold-adapted Rupi-caprini. The musk ox is a remnant of a radiation of diverse giant caprids that arose in the Villafranchian, paralleling a similar radiation in the Old World deer. These early caprid giants included not only the lineage of musk oxen, but also that of the euceratheres that survived in North America virtually to the Recent, and odd genera such as *Soergelia*. We also find small-bodied forms in the early caprid radiation such as *Myotragus*, *Tossunnoria*, *Sivicapra*, *Pachygazella*, *Procamptoceras* and *Gallogoral*.

It is evident from Figure 1.1 that the caprid radiation evolved an abundance of grotesque giants from alpine, arctic and periglacial environments, but also one from the hot desert, *Ammotragus*. We find repeatedly the concomitant differentiation of head and rear pole with evolutionary advancement, and in northern forms a tendency for reduced tail size and increased rump patch size. Tails are likely to freeze in very cold weather. Nor are the island dwarfs missing. They are illustrated by the small Taiwanese and Japanese serows, and by the tiny cave-goat (*Myotragus*) from Mallorca. In North America bighorns and mountain goats penetrated into the southern parts of the continent, paralleling a similar advance by other mammals such as elk, moose, grizzly bears, wolves and modern bison. Mountain goat and sheep show a reduction in body size (Harris & Mundel, 1974; Kurten & Anderson, 1980) just as did the wapiti (*Cervus canadensis*) (Guthrie, 1966). The extinct southern mountain goat has even been granted specific status (*Oreamnos harringtoni*); it penetrated to Mexico during the Wisconsian maximum and co-existed with the Rancho-La Brean fauna (Kurten & Anderson, 1980). Some recent unpublished findings by George T. Jefferson of the Natural History Museum, Los Angeles, indicate that *Ovis* may have had a continuous existence in the area of the Mohave Desert in California since 300,000 BP. A second but Wisconsian area of bighorn survival appears to be in Wyoming where a more or less continuous record reaches to the Sangamon interglacial (Miles B. Gilbert, personal communication). Bighorns, however, did not make headway against the indigenous Rancho-La Brean fauna of giant herbivores and carnivores till about 10,000-17,000 BP. Thereafter they spread radially from the Wyoming centre and occupied most of their present range well before the change to the modern climates, and well before the demise of the Rancho-La Brean fauna (Geist, in preparation).

The evolutionary model described can also be used to unravel tangled relationships such as those within *Ovis* and *Capra*. Note that the model predicts that new forms arise peripherally to parental forms, and occupy climatically more extreme areas, that such areas be of large size and that the new forms have enhanced head and rear poles. These criteria would make *Ammotragus* a descendant of *Hemitragus*; no fossil evidence exists to place *Ammotragus* in time, but the model predicts that it arose probably early in the Villafranchian period.

I have suggested earlier (Geist, 1971a) that both *Capra* and *Ovis* arose from *Ammotragus*; that it carries a mixture of sheep and goat-like traits is agreed and in no need of review (see Schaller, 1977). To be descendants, both *Capra* and *Ovis* must fulfil the conditions enumerated above. They do so. They represent a step from warm to cold desert conditions, are adjacent in distribution to *Ammotragus* and continue the typical caprid trend of increasing horn size while reducing long display hair. The heteronym hornwinding, because it is an *Ammotragus* trait, is considered primitive.

The ranges of *Capra* and *Ovis*, however, have been run over by many glacial advances and concomitant floristic and climatic changes (see Nilsson, 1983); we expect extinctions, reinvasions and speciation. We expect some hybridisation, but we do not expect in a herbivore displacement of an older form by a younger form. Although *Capra* gravitated to cliffs and browsing, and *Ovis* gravitated more to rolling terrain and grazing so that their ecological requirements are somewhat different, we still expect that when *Ovis* disperses so does *Capra* even though the spatial overlap need not coincide. If so, we expect a parallel between *Ovis* and *Capra* species in that primitive sheep will be geographically with primitive goats, and evolved sheep will be with evolved goats. Recent zoogeographic radiations are more likely to be contiguous than earlier ones; the latter are more likely to be fragmented and reduced. Also, episodes of major extensive glaciations will likely lead to greater zoogeographic dispersal of caprids and cold-adapted rupicaprids than will minor glacial pulses.

Given the foregoing consideration we note at least four radiations of sheep and goats.

1. The oldest advance is represented today by the 2n=58 chromosome urials, with bibs and small, diffuse rump patches

(Ladak, Punjab, Afghan and Transcaspian urials), and the markhor. The latter is relatively primitive in dentition, has a haircoat which in males is reminiscent of *Ammotragus*, has horns showing an extreme form of heteronym winding, and is adapted to warmer, less-extreme climates than are the other goats. It is a lower altitude browser on steep slopes and cliffs, just as urials are basically low-altitude sheep compared to argalis. The distribution of these urials and the markhor roughly coincides being pinched in between those of the other sheep-goat pairs to be described, and being the smallest in geographic extent.

2. The next oldest radiation is represented by the ruffed argalis (the Tibetan argalis) and the 'round-horned goats', *Pseudois* in particular. The Tibetan argali is primitive by the criterion that horns replace ruffs in the evolutionary history of *Ovis*. Its distribution coincides with that of *Pseudois*, a goat very similar to pachycerine sheep in ecology, which means that it is bound to cliffs for escape terrain, but is not as extreme in its adaptations to cliffs as is the ibex. The distribution of *Pseudois* appears to be the larger; the Tibetan argali is found within it, restricted more to xeric areas. *Pseudois* also appears to overlap with the Mongolian argalis, which in part have lost their bibs. The ruffed argali and *Pseudois* are segregated ecologically via adaptation to high elevation from the abutting urials and markhor.

To the west of the Tibetan plateau, we have no argali radiation although 'argali'-like sheep did appear in mid-Pleistocene times in Europe (Herre & Kesper, 1953; Korobitsyna *et al.*, 1974). We do have, however, remnants of heteronym-horned goats, such as the east Caucasian tur, and in the extreme west the Spanish ibex. The latter may well be a product of hybridisation of several *Capra* advances. That some hybridisation can and does occur is indicated by the hybrid zone between east and west Caucasian turs (Heptner, Nasimovisch & Bannikov, 1961), the unlimited fertility of *Capra* hybrids, and the account by Heptner *et al.* (1961) of the breakdown of reproduction barriers where turs and wild goats were introduced to vacant ranges simultaneously.

If one could date *Pseudois* via fossil remains, one could probably date this advance. It is likely as far back as the middle of the major glaciations.

3. The next advance is represented by the 2n=54/52 chromosome sheep, be they with or without ruff, and the wild goat

(*Capra aegagrus*). This was a massive radiation and probably included the ancestor of *Rupicapra* and *Oreamnos* as well. It was probably associated with the largest of the major set of glacial advances, maybe with the penultimate or Riss glaciations, about 300,000 years BP (Rutter, 1980).

The 2n=54 chromosome sheep comprise the moufflons and the pachycerine sheep, which are exceedingly similar in karyotype (Nadler *et al.*, 1974; Korobitsyna *et al.*, 1974); in Siberia the pachycerine snow sheep (*Ovis nivicola*) have 52 chromosomes (Korobitsyna *et al.*, 1974). The 2n=54 sheep are at the extremes of sheep distribution having reached Europe and Africa on one hand, and North America on the other. The moufflons carry no bib, but do have a neck ruff, albeit a reduced one. They have, like pachycerine sheep, flat tails and distinct rump patches. Otherwise they are very urial-like. On their eastern flank, they form a massive hybrid zone with urials (Valdez, Nadler & Bunch, 1978). These are probably the small sheep that appear in late Pleistocene times in Europe.

The pachycerine or American-type cold-adapted sheep appear in pre-Illinoian (Riss) deposits in California and in Illinoian (late Riss) deposits of Alaska (Kurten & Anderson, 1980). They colonised the south late in the Wisconsin glaciation prior to megafaunal extinction, but became abundant only after the collapse of the Rancho-La Brean fauna post-glacially (Geist, in preparation). Like moufflons, they have distinct rump patches and broad tails. Their biology, however, is shaped both by the absence of *Capra* competition and life in cold, long winters (Geist, 1971a). Sheep could only have reached eastern Siberia during some massive glaciations; pachycerine sheep are glacier followers *par excellence*. So is the mountain goat (*Oreamnos*) and the chamois (*Rupicapra*).

The wild goat is advanced by virtue of its more molarised, subhypsodont dentition. It has greatly diverged in appearance from earlier goats and its horns have changed to a type almost void of heteronym winding. The Chiltan goat, appropriately located close to markhor and the normal wild goat, does show heteronym winding in its horns (Schaller, 1977). The distribution of wild goats is confined largely to Asia Minor and the east, but it may have been present in Europe according to Schaller (1977). Unfortunately, one also may be dealing here with feral goats.

4. The most recent advance is that of ibex and the bibless

argalis. The ibex races show a radial distribution about that of the wild goat. The distribution is quite contiguous despite its great size. Ibex are closely related to wild goats, but are advanced evolutionarily by our criteria. They also occupy climatically more extreme environments than the wild goat (see Nievergelt, 1981). The only sheep radiation associated with that of the ibex is the argali radiation that roughly parallels that of the Siberian ibex. As in the case of the ibex, the bibless argalis are highly evolved and form a cline running from the Karka Tau desert to the Altai in which horn mass enlarges with body size, and the vestiges of the bib are lost. They have a karyotype of 2n=56 chromosomes and are thus very close to the 2n=58 chromosome urials, from which I suspect they descended (Geist, 1971a). The ibex advance can be dated; it falls late into the Riss glaciations. Ibex are numerous in Europe, along with the chamois, in the following interglacial, and the last major glacial. These two species indicated that they advanced under mesic conditions which favoured shrubs and forests, excluding the sheep. These do show up in the mediterranean area including North Africa and Italy (Kurten, 1968), as well as in the Caucasus (Korobitsyna *et al.*, 1974).

The foregoing is a brief sketch that poses a good many questions. I shall leave it at that.

References

Bailey, J.A. (1980) 'Desert Bighorn, Forage Competition, and Zoogeography', *Wildl. Soc. Bull.*, *8*, 208-16

Beninde, J. (1937) *Zur Naturgeschichte des Rothirsches*, Monog. Wildsaugetiere 4 P. Schops, Leipzig

Churcher, C.S. (1978) 'Giraffidae' in J. Maglio & Cook H.B.S. (eds.), *Evolution of African Mammals*, Harvard University Press, Cambridge, pp. 509-35

Geist, V. (1971a) *Mountain Sheep*, Chicago University Press, Chicago

—— (1971b) 'On the Relation of Social Evolution and Dispersal in Ungulates During the Pleistocene, with Emphasis on the Old-World Deer and the Genus *Bison*', *Quaternary Research*, *1*, 283-315

—— (1978) *Life Strategies, Human Evolution and Environmental Design*, Springer-Verlag, New York

—— (1983) 'On the Evolution of Ice Age Mammals and Its Significance to an Understanding of Speciation', *Assoc. South-East Biol. Bull.*, *30*, 109-33

Gingerich, P.D. (1977) 'Patterns of Evolution in the Mammalian Fossil Record' in A. Hallum (ed.) *Patterns of Evolution as Illustrated by the Fossil Record*, Elsevier, Amsterdam, pp. 469-500

Gould, J.S. & Eldredge, N. (1978) 'Punctuated Equilibria. The Tempo and Mode of Evolution Reconsidered', *Paleobiology*, *3*, 115-51

Guthrie, R.D. (1966) 'The Extinct Wapiti of Alaska and the Yukon Territory', *Can. J. Zool*, *44*, 47-57

Harris A.H. & Mundel, P. (1974) 'Size Reduction in Bighorn Sheep (*Ovis canadensis*) at the Close of the Pleistocene', *J. Mammal.*, *55*, 678-80

Heptner N.G., Nasimovisch, A.A. & Bannikov, A.G. (1961) *Die Saugetiere der Sowjetunion*. Band 1: Paarhufer und Unpaarhufer. V.E.B. Gustav Fischer Verlag, Jena (1966)

Herre W. & Kesper, K (1953) 'Zur Verbreitung von *Ovis ammon* in Europe', *Zool. Anz.*, *151*, 204-9

Korobitsyna, K., Nadler, C., Vorontsov, N. & Hoffman, R. (1974) 'Chromosomes of the Siberian Snow Sheep, *Ovis nivicola*, and Implications Concerning the Origin of Amphiberinglau Wild Sheep (Subgenus *Pachyceros*)', *Quater. Res.*, *4*, 235-45

Kuck, L. (1977) 'The Impact of Hunting on Idaho's Pahsimeroi Mountain Goat Herd' in W. Samuel & Macgregor W.G. (eds.) *Proc First International Mountain Goat Symposium*. Province of British Columbia, Ministry of Recreation and Conservation. Victoria, BC, pp. 114-25

Kurten, B. (1968) *Pleistocene Mammals of Europe*, Weidenfeld & Nicolson, London

Kurten, B. & Anderson, E. (1980) *Pleistocene Mammals of North America*, Columbia University Press, New York

Margalef, R. (1963) 'On Certain Unifying Principles in Ecology', *Am. Natur.*, *97*, 357-74

Nadler, C. Hoffman, R. & Woolf, A. (1974) 'G-band Patterns, Chromosome Homologies and Evolutionary Relationships Among Wild Sheep, Goats and Aoudad (Mammalia, Artiodactyla)', *Experientia*, *30*, 744-46

Nievergelt, B. (1981) *Ibexes in an African Environment*, Springer Verlag, New York

Nilsson, T. (1983) *The Pleistocene*, Reidel, Dordrecht

Rutter, N.W. (1980) 'Late Pleistocene History of the Western Canadian Ice-free Corridor', *Can. J. Anthropol.*, *1*, 1-8

Schaller, G.B. (1977) *Mountain Monarchs*, Chicago University Press, Chicago

Stock, A. & Stokes, W. (1969) 'A Re-evaluation of Pleistocene Bighorn Sheep from the Great Basin and Their Relationship to Living Members of the Genus *Ovis*, *J. Paleontol.*, *50*, 805-7

Stokes, W.M.L. & Condie, K.C. (1961) 'Pleistocene Bighorn Sheep from the Great Basin', *J. Paleontol.*, *35*, 598-609

Tchernov, E. (1981) 'The Impact of the Postglacial on the Fauna of Southwest Asia' in W. Frey & H.P. Verpmann (eds.), *Beitrage zur Umweltgeschichte des Vorderen Orients.* Beiheft zum Tubinger Atlas des Vordern Orients, Reihe A. No. 8, pp.197-216

Thenius, E. & Hofer, H. (1960) *Stammesgeschichte der Saugetiere. Eine Ubersicht über Tatsachen und Probleme der Evolution der Säugetiere*. Springer Verlag, Berlin

Valdez R., Nadler, C.F. & Bunch, T.D. (1978) 'Evolution of Wild Sheep in Iran', *Evolution*, *32*, 56-76

Vogt, F. (1936) *Neue Wege der Hege*, Verlag Neumann, Neudamm

CHAPTER 2

Würmian and Holocene Chamois of Italy

F. MASINI

Dipartimento di Scienze della Terra, Universitá di Firenze, Via La Pira 4, 50121 Firenze (Italy)

Abstract

Remains of *Rupicapra rupicapra* from five Würmian and Holocene sites of Italy have been studied. The taxonomy and distribution of the Pleistocene chamois in Western Europe, with particular reference to the fossils of Italy, are briefly discussed.

R.r.rupicapra is present in the Würm 1 of the eastern Alps and in the Würm 2 of the Maritime Alps as well as of the northern Apennines, where it survived up to the Holocene.

Up to now, fossils of *R.r.ornata* have been collected only from Late glacial and Holocene layers of the central and southern Apennines.

Recent taxonomic studies have confirmed the hypothesis that *R.r.ornata* and *R.r.pyrenaica* are closely related forms and suggest that they originate from a common ancestor in the Late Pleistocene.

The migration of *R.r.rupicapra* into the western regions marks the subdivision of the ancient *pyrenaica-ornata* geographic range. On the basis of palaeontological data the immigration may have occurred at least at the beginning of the last glacial period.

Introduction

Recent electrophoretic work carried out on *R.r.pyrenaica*, *R.r.ornata* and *R.r.rupicapra* confirms the hypothesis that the two former subspecies should be considered phylogenetically more closely related to each other than to *R.r.rupicapra*. The genetic flow between them must have been interrupted in the

Würm (Nascetti *et al.*, 1985). The interruption was probably due to the immigration of the Alpine form (*R.r.rupicapra*) which occupied a part of the *pyrenaica-ornata* geographic range (Lovari & Scala, 1980). The purpose of this study is to determine the *rupicapra* immigration date on the basis of palaeontological data.

The best diagnostic characters of the skeleton for distinguishing the *pyrenaica-ornata* forms from the Alpine subspecies are the obliterated ethmoidal vacuity, the complexity and tightness of the skull sutures and the shape of the frontoparietal suture (Camerano 1913-14-15; Couturier, 1938). The morphology of the horn cores is useful for distinguishing *pyrenaica* from *ornata* (Couturier,1938) but can hardly be utilised to recognise their common ancestor, as it differs considerably in the two living subspecies (Masini, in preparation).

Unfortunately, fossils from the Pleistocene deposits usually consist of fragmentary remains of which only the horn cores can be used in taxonomy; this makes it difficult to classify the fossil records.

I am grateful to S. Lovari, University of Parma, for his constructive discussions, and to all those who supplied me with the material studied. I am particularly grateful to F. Pedrotti, University of Camerino, for the material of Ripa Grande; E. Borzatti von Löwenstern, University of Florence, for the Papasidero material; R. Fondi, University of Siena, for the metric data on the living forms; B. Sala, University of Ferrara, for the material of Grotta del Brojon; the Ente Parco Nazionale d'Abruzzo, for the skulls of the *ornata* subspecies.

Data on the Distribution and Taxonomy of the Fossils from Western Europe

Little information is available on the distribution of fossil forms and even scantier data are reported about their taxonomy.

An outline of chamois distribution in western Europe is shown in Figure 2.1. The most ancient remains come from the Caune de l'Arago (Eastern Pyrenees) from layers which have been ascribed either to the end of the Mindel (Cregut & Guèrin, 1979) or to Riss 1 (Chaline, 1977).

Rare and fragmentary remains are present in the Rissian deposits of French territory (Delpech & Heintz, 1976; Guèrin,

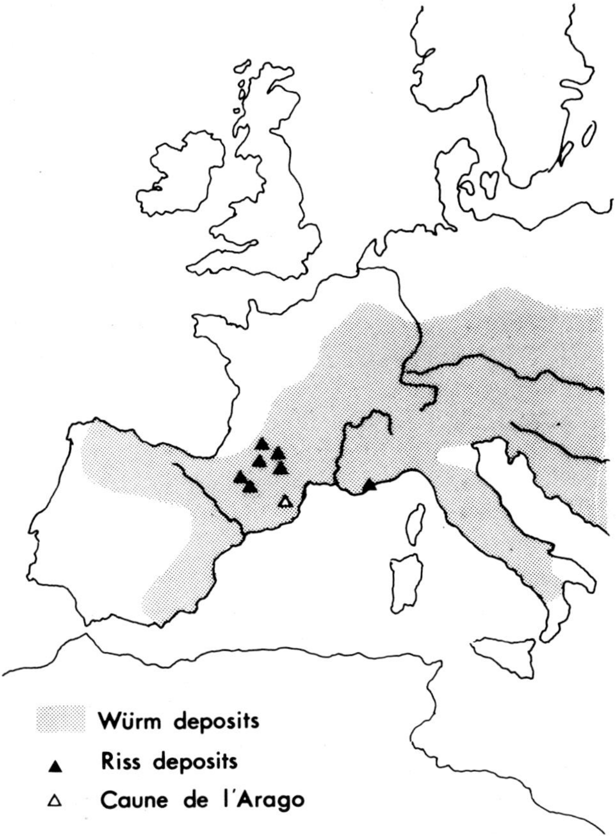

Figure 2.1: Outline of *Rupicapra* Distribution in the Western European Pleistocene. From literature data

1980). Finds from the Riss-Würm interglacial seem to occur only in two localities of the Swiss Alps (Dubois & Stehlin, 1933). The literature lacks any information about the taxonomy of these pre-Würmian forms.

During the Würm and Holocene finds of *Rupicapra* are more numerous, demonstrating its very large distribution. These remains come from deposits related to the main mountain ranges and to locations of minor isolated relief as well.

Forms with horn cores of *pyrenaica* type have occurred since the older Würm on the Cantabrian Mountains (Altuna, 1971) and on the Pyrenees (Koby, 1964), while the existence of the

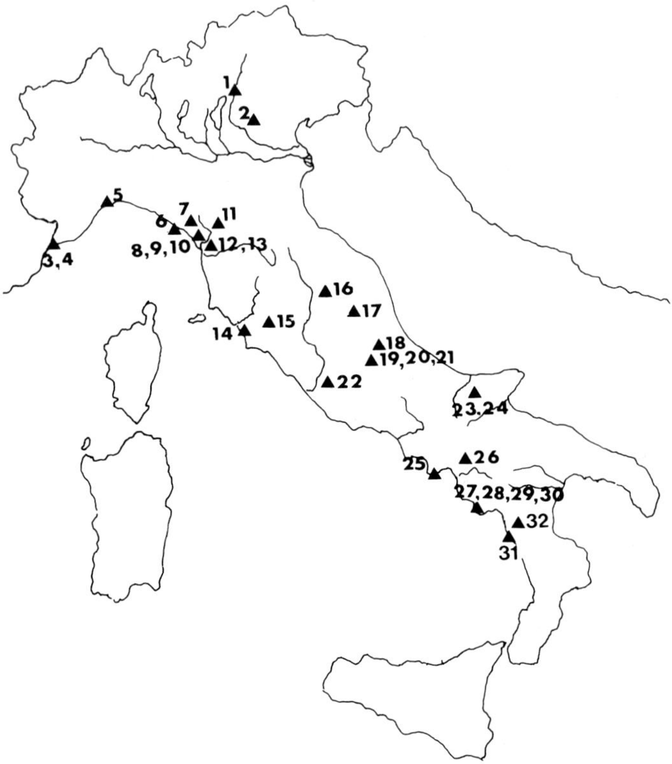

Figure 2.2: Geographical Distribution of Some Italian Sites in which *Rupicapra* Remains are Present. The numbers refer to references in Table 2.1

rupicapra subspecies in the Würm 4 of the Ardennes is demonstrated by a well-preserved skull (Boule, 1910).

The distribution and age of deposits in the Italian peninsula, where chamois remains were found, are reported in Figure 2.2 and Table 2.1. Boule (1910) refers a horn core from the Grotta del Principe to a *pyrenaica*-like form and two horn cores from the Grotta dei Fanciulli to *rupicapra*. Camerano (1913-14-15) assigns the horn cores from Equi to the latter subspecies.

Table 2.1: Age of the Deposits Reported in Figure 2.2. Bibliographical references are given in Masini (1983)

1. Ripari di valle dell'Adige: Holocene; (Boscato & Sala, 1980)
2. G. del Brojon: Riss (?), W.1 – W.4; (Sala, 1980)
3. G. del Principe: W.1 – W.1; (De Lumley & Woodyear, 1969)
4. G. dei Fanciulli: W.3 – W.4; (Palma di Cesnola, 1976)
5. G. di S. Lucia: Würm 2; (Tozzi, 1962)
6. G. dei Colombi: W.3 – W.4 (?); (Regalia, 1894)
7. G. di Equi: Würm 2; (Pitti & Tozzi, 1971)
8. G. all'Onda: W.2 – W.2; (Pitti & Tozzi, 1971)
9. Buca del Tasso: Würm 2; (Pitti & Tozzi, 1971)
10. M. Sumbra: Holocene; (Masini, 1983)
11. Riparo delle Campane: Late Glacial; (Biagi *et al.*, 1981)
12. G. di Parignana: Würm (undet.); (Forsyth Major, 1872)
13. G. di Cucigliana: Würm (undet.); (Mochi, 1911)
14. G. la Fabbrica: W.3 – W.4; (Pitti *et al.*, 1976)
15. G. di Gosto: Würm 2; (Tozzi, 1974)
16. Monte Cucco: Würm (undet.); (Piazza, 1982)
17. Ripa Grande: Holocene (Masini *et al.*, in preparation)
18. Riparo I Grottoni: Würm 1 (?); (Giustizia, 1976)
19. G. di Ortucchio: Younger Würm; (Cremonesi, 1968)
20. G. di S. Nicola: Würm 4; (Borzatti von Löwenstern, 1962)
21. G. Achille Graziani: Würm 3; (Radmilli, 1955)
22. G. Polesini: Late Glacial; (Cassoli, 1977)
23. G. Spagnoli B: Würm 1; (Sala, 1978)
24. G. Paglicci: W.3 – W.4; (Palma di Cesnola, 1975)
25. G. Erica: Holocene; (Bonuccelli, 1970)
26. G. di Castelcivita: Würm 3; (Cioni *et al.*, 1979)
27. Riparo del Poggio; Riss (?); (Sala, 1979)
28. G. Tina: Würm 2; (Martini *et al.*, 1972-74)
29. G. della Cala: W.4 – Late Glacial; (Bartolomei *et al.*, 1976)
30. G. Calanca: Younger Würm; (Vigliardi, 1968)
31. G. di Torre Nave: Würm 4; (Bulgarelli, 1972)
32. Papasidero: Late glacial – Holocene; (Bulgarelli, 1972)

Material and Methods

The material considered in this paper consists of the following specimens:

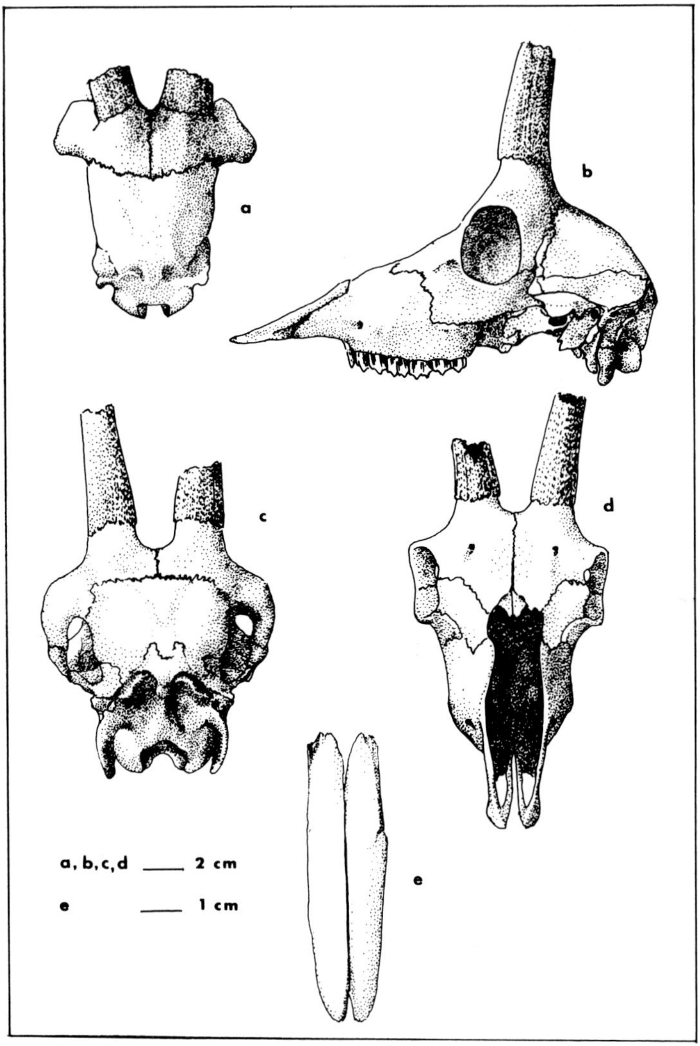

Figure 2.3: Monte Sumbra Specimens. a: skull IGF 1277V, dorso caudal view; b: skull IGF 1259V, lateral view; c: skull IGF 1259V, caudal view; d: skull IGF 1259V, dorso frontal view; e: nasal bones IGF 1260V, dorsal view. The letters IGF indicate that the material is conserved in the Museo di Geologia e Paleontologia dell'Università di Firenze

M.Sumbra (Apuan Alps, Holocene): three adult male skulls; one is very well preserved (Figure 2.3).
Equi (Apuan Alps, Würm 2): 32 isolated horn cores of both male and female individuals and of different ages.
Grotta del Brojon (Berici Hills, Würm 1): four horn cores, two very incomplete ones.
Papasidero (southern Apennines, Late Glacial — Holocene): two horn cores; one belongs to a young individual.
Ripa Grande (M.Sibillini, Holocene): two incomplete female skulls; one belongs to a young individual (Masini *et al.*, in preparation).

These data are summarised from a more extensive report (Masini, 1983) except for Ripa Grande.

This material was compared with the living *rupicapra*, *ornata* and *pyrenaica* subspecies. The taxonomic analysis evaluated the morphological features of the skull as well as details and metric parameters of horn cores.

The taxonomic features of the skull pointed out by Camerano (1913-14-15) and Couturier (1938) were critically revised by comparing 18 skulls of *rupicapra* with 23 skulls of *ornata*. Metric data, both original and from Fondi (unpublished) were integrated with the information available in Camerano (1913-14-15) and Couturier (1938).

Results

The features of the skulls of *pyrenaica*, *ornata*, *rupicapra* and those of the skulls from M.Sumbra are compared in Table 2.2; those relevant to horn cores in Table 2.3. The differences between the living forms are not pronounced and their features are often irregularly distributed among the subspecies. Most of these characters, taken singly, do not permit a full discrimination, but their frequency and association allow a fairly good taxonomic assessment.

The value of the elementary statistics of some horn core parameters for the living forms, M.Sumbra and Equi specimens are reported in Table 2.4.

Figure 2.4 shows the regression lines of minimum distance between horn cores on their antero-posterior diameter for the

Characters	*pyrenaica*	*ornata*	*rupicapra*	M.Sumbra
1. Ethmoidal vacuity	absent	absent	usually present	present
2. Shape of nasal bone	a. oval b. triangular, without lachrymal apophysis	b. triangular, without lachrymal apophysis	a. oval c. triangular, with lachrymal apophysis	a. oval
3. Anterior notch of frontals	usually narrow	wide	narrow	narrow
4. Length of the medial edge of lachrymal bone	medium	great	small	medium (inferred)
5. Frontonasal suture	irregular	irregular	regular	regular
6. Caudal notch of palatines	'U' shaped	rounded, with small median incisure	usually 'V' shaped	'V' shaped
7. Zygomatic process of frontals	wide	narrow	wide	wide
8. Caudal opening of the orbits	small	large	small	small
9. Frontoparietal suture	with a straight median part much indented	with a straight median part much indented	forming an obtuse angle open anteriorly less indented	forming an obtuse angle open anteriorly less indented
10. Skull bones sutures	closed and tending to be complex	closed and tending to be complex	more open and simple	more open and simple
11. Postcornual fossettes	(?)	pronounced	less pronounced	less pronounced
12. Frontal bone	little raised at the horn core bases	raised at the horn core bases	raised at the horn core bases	raised at the horn core bases

Table 2.3: Horn Core Characters in *R.r.pyrenaica*, *R.r.ornata*, *R.r.rupicapra* and in Monte Sumbra Specimens

Characters	*pyrenaica*	*ornata*	*rupicapra*	M.Sumbra
1. Insertion	close	close	less close	less close
2. Divergence	m+f:sub-parallel	m:divergent f:sub-parallel	m:divergent f:sub-parallel	m:divergent
3. Medio-lateral course	m+f:almost straight	m:curved outward f:almost straight	m:curved outward f:almost straight	m:curved outward
4. Antero-posterior inclination	m+f:anteriorly inclined	m+f:posteriorly inclined	m+f:anteriorly inclined	m:posteriorly inclined
5. Dimensions	among the smallest of the species	among the greatest of the species	great dimensional variation	medium
6. Other characters:	in *ornata* the following characters are often present: a. sub-triangular section with caudal base b. irregular medio-lateral constrictions in adult and old individuals c. tendency to develop an anterior keel in the distal half (mostly in adult males) d. strong medio-lateral compression.			

m = males; f = females.

Table 2.4: Basic Descriptive Statistics of Some Horn Core Variables in Living and Fossil Samples

	N	Min	Max	$\bar{X}$	SD	SE
R.r.pyrenaica (m + f)						
APD	18	12.5	22	17.41	2.29	0.12
TD	18	10.5	19.5	15.5	2.21	0.52
MD	18	8	15	12.3	2.36	0.55
CNA	17	92°	111°	102°5′	5°00′	1°22′
R.r.ornata (m + f)						
APD	32	19.8	29	24.8	3.32	0.58
TD	32	17.7	26	21.22	3.18	0.69
MD	21	9.5	15.8	12.7	1.93	0.44
CNA	14	104°	120°	111°97′	4°30′	1°12′
R.r.rupicapra (m + f)						
APD	135	15	27	20.9	3.43	0.29
TD	139	12	26	18.44	3.56	0.30
MD	132	9	24	16.2	2.59	0.22
CNA	50	90°	114°	95°36′	5°48′	0°48′
M. Sumbra (m)						
APD	3	22.8	26	24.46	1.60	0.92
TD	3	20	22.2	20.73	1.27	0.73
MD	3	14.5	19.0	16.83	2.25	1.30
CNA	1	115°	115°	115°	/	/
Equi (m + f)						
APD	31	15	23.6	19.2	2.70	0.74
TD	28	12.1	20.2	16.6	2.49	0.78
MD	13	12	19	14.9	2.26	0.62

Explanations: N = number of specimens; Min = minimum value observed; Max = greatest value observed ; $\bar{X}$ = mean; SD = standard deviation; SE = standard error of the mean; APD = antero-posterior diameter of horn core; TD = transverse diameter of horn core; MD = minimum distance between horn core bases; CNA = cerato-nasal angle; m = males; f = females.

living forms and for the Equi sample. The living forms follow different regression lines. In particular, *pyrenaica* separates well from the other subspecies.

The skulls of M.Sumbra show almost all characters of the

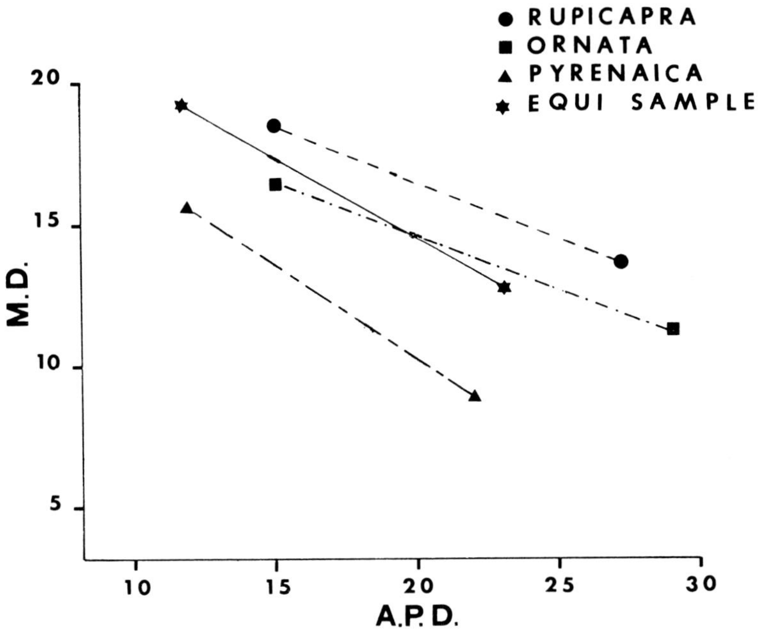

Figure 2.4: Linear Regression of Minimum Distance between Horn Core Bases (MD) on Antero-posterior Diameter of the Horn Core (APD) for the Living Subspecies Considered in this Paper and for the Equi Sample

living Alpine form (cf. Tables 2.2, 2.3 and Figure 2.3) from which they differ in the backwards inclination of the horn cores, as in *ornata*.

On the other hand the features of the Ripa Grande skulls are typical of *ornata* (Masini *et al.*, in preparation).

The adult Papasidero horn core shows a strong medio-lateral compression, a sub-triangular section and a narrow anterior edge: thus, it can be ascribed to *ornata*, too.

The Equi specimens are medium-small and many show a marked lateral bending which makes them different from *pyrenaica*. Unlike *ornata* they lack medio-lateral constrictions, anterior keel and sub-triangular section. The inferred minimum distance between horn cores shows intermediate values between *rupicapra* and *ornata*, but it differs significantly (n=32, t=2.815, p<0.01) from the latter. Although the regression line of horn core parameters is somewhat steeper than the *rupicapra* one and closer to that of *pyrenaica* (cf. Figure 2.4), all other features of the Equi sample fit the *rupicapra* form to which it may be referred.

The main feature of the Grotta del Brojon remains is the low longitudinal growth. The characters of this sample allow it to be included in the morphological range of *rupicapra*.

Discussion and Conclusions

A tentative outline of distribution of the chamois subspecies considered in this paper, during the Würm and Holocene, is shown in Figure 2.5.

The Papasidero and Ripa Grande specimens prove that *ornata* was already differentiated at least in the Würm 4 (about 15,000 years BP) and that up to the early Holocene it had a much wider

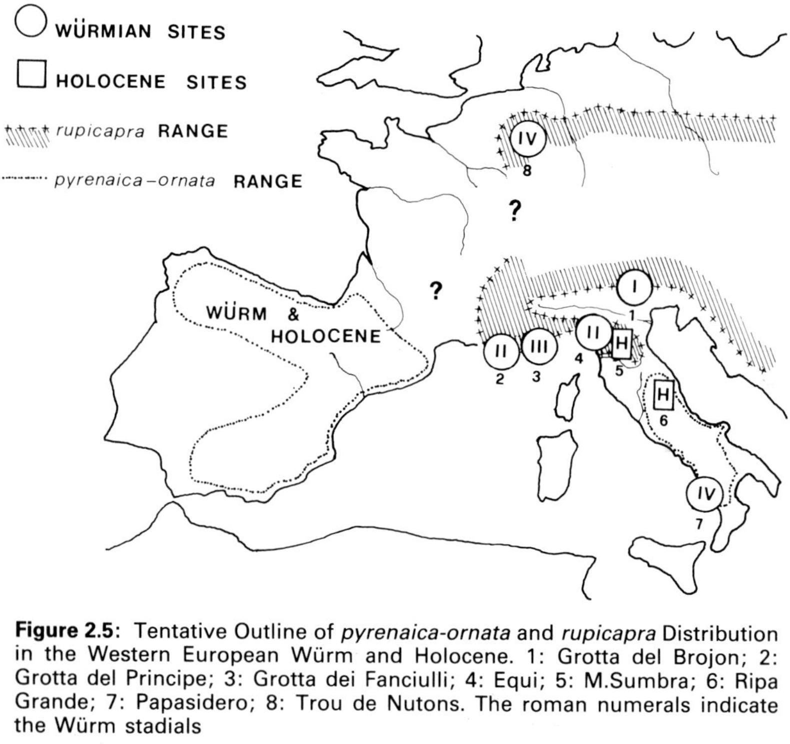

Figure 2.5: Tentative Outline of *pyrenaica-ornata* and *rupicapra* Distribution in the Western European Würm and Holocene. 1: Grotta del Brojon; 2: Grotta del Principe; 3: Grotta dei Fanciulli; 4: Equi; 5: M.Sumbra; 6: Ripa Grande; 7: Papasidero; 8: Trou de Nutons. The roman numerals indicate the Würm stadials

range than its supposed range in historical times (cf. Lovari & Scala, 1980).

The Equi specimens can be referred to a form similar to the living *rupicapra*. Contrary to the interpretation of Boule (1910), the horn core collected in the Grotta del Principe could also be ascribed to an Alpine chamois, because of its morphological likeness to the Equi specimens, the geographical contiguity and equivalent age of the two localities.

Also the Würm 3 horn cores of the Grotta dei Fanciulli and the Holocene remains of M.Sumbra belong to *rupicapra*.

As *rupicapra* has been present in the eastern Alps since the Würm 1 (Brojon) and during the Würm 2 in the northern Apennines as well (Equi), it may be concluded that this chamois is likely to have migrated along the Alpine arch at least at the beginning of the last glaciation, thus dividing the geographical range of *pyrenaica-ornata* in the last stage of its westward migration. Such results support the hypothesis, proposed by Lovari & Scala (1980), about chamois radiation in south-western Europe.

At present, a major limitation to the interpretation of these data is the absence of taxonomical references relevant to the pre-Würmian forms, as well as the very scanty data about the taxonomy and distribution of the eastern European and Asiatic fossil forms. A more extensive and accurate study will be necessary to acquire a less partial picture of chamois evolution.

Furthermore, the study of body size of fossils would be interesting in order to assess the possibility of significant changes during the Würm and Holocene. The data from Guipùzcoa (Cantabrian Mountains) prove that the small body size of the living *R.r.parva* is a recently acquired feature, as the fossil chamois of Cantabrian sites was almost the same size as the living Alpine form (Altuna, 1971). Preliminary observations concerning the Papasidero limb bone remains also suggest that these late glacial representatives of *ornata* were larger than the living ones. On the other hand the Holocene Ripa Grande remains are almost of the same size as the living *ornata*.

A particular problem is represented by the very small genetic and phenotypical variability of the living *ornata* (Nascetti *et al.*, 1985; Masini, 1983). This subspecies passed through a recent bottle-neck, which occurred during World War II (Lovari & Perco, 1980), and now is represented by a small homozygotic population living in a restricted area of the Central Apennines. Therefore the low morphological variability of such a population probably does not allow a good knowledge of the features of the original *ornata* subspecies. The data relevant to the morphology of the fossil representatives of *ornata* would help to evaluate better the taxonomic significance of the osteologic features of this subspecies.

References

Altuna, J. (1971) 'Fauna de Mamíferos de los Yacimientos Prehistoricos de Guipùzcoa', *Munibe, 1972*, 1-464

Boule, M. (1910) 'Paléontologie' in Villeneuve, Boule, Vernan, Carthailac (eds.), *Les Grottes de Grimaldi (Baousse Roussé)*, vol. 1 pp. 157-236

Camerano, L. (1913-14-15) 'Ricerche Intorno ai Camosci: Parte 1a, 2a, 3a'. *Mem. R. Acc. Sci. Torino (Cl. Sci. Mat. Fis. Nat.), S. II. 64*, 1-82; *64*, 1-88; *65*, 1-82

Chaline, J. (1977) 'Essai de Biostratigraphie et de Correlations Climatiques du Pléistocène Inférieur et Moyen Continental Holartique d'après l'Évolution et la Dynamique des Migration de Rongeur', *Suppl. Bull. AFEQ*, *50*, 349-61

Couturier, M. (1938) *Le Chamois*. Arthaud, Grenoble

Cregut, E. & Guèrin, C. (1979) 'Première Découverte en Europe Sud-occidentale de *Praeovibos priscus* (Mammalia, Artiodactyla, Ovibovinae) dans le Gisement Pléistocène Moyen Ante-rissien de la Caune de l'Arago (Tautavel, Pyrénées orientales, France)', *Geobios*, *12*, 459-65

Delpech, F. & Heintz, E. (1976) 'Les Artiodactyles: Bovidés' in H. de Lumley (ed.), *La Prehistoire Francaise* vol. 1, pp. 286-94

Dubois, A. & Stehlin, H.G. (1933) 'La grotte de Cotencher', *Mem. Soc. Paleont. Suisse*, *52*, 133-6

Guèrin, C. (1980) 'Les Rhinoceros (Mammalia, Perissodactyla) du Miocène Terminal au Pléistocène Supérieur en Europe Occidentale', *Doc. Lab. Sci. Lyon*, *79*, 1-1185

Koby, F. (1964) 'Ostéologie de *Rupicapra pyrenaica* d'après les Restes de la Caverne de la Vache', *Bull. Soc. Prehist. Ariege*, *19*, 13-31

Lovari, S. & Perco, F. (1980) *Il camoscio d'Abruzzo*. Doc. Parco Naz. Abruzzo, Ente Autonomo P.N.A., Pescasseroli (AQ)

Lovari, S. & Perco, F. (1980) *Il Camoscio d'Abruzzo*. Doc. Parco Naz. Abruzzo, evaluation of Couturier's Data on the Morphometry of Six Chamois Subspecies', *Boll. Zool.*, *47*, 113-24

Masini, F. (1983) 'I Camosci Sub-fossili del M.Sumbra (Alpi Apuane)', Tesi di laurea in Sci. Biol. A.A. 1981-82, Bibl. Dip. Geominer. Univ. Firenze

Nascetti, G., Lovari, S., Lanfranchi, P., Berducou, C., Mattiucci, S., Rossi, L. & Bullini, L. (1985) 'Revision of *Rupicapra* Genus. III. Electrophoretic Studies Demonstrating Species Distinction of Chamois Populations of the Alps from those of the Apennines and Pyrenees', in Lovari, S. (ed.), *The Biology and Management of Mountain Ungulates*, Croom Helm, London, pp. 56-62

CHAPTER 3

Reclassification of the Serows and Gorals (*Nemorhaedus*: Bovidae)

C.P. GROVES[1] and P.GRUBB[2]

1. Department of Prehistory and Anthropology, Australian National University, Post Office Box 4, Canberra ACT 2600 (Australia)
2. 35 Downhills Park Road, London N17 6PE (UK)

Abstract

Studies of museum specimens of serows and gorals suggest that the two groups should be assigned to one genus, *Nemorhaedus*. Of serows, there are three species, *N. sumatraensis*, *N. swinhoei*, and *N. crispus*. The more specialised gorals include three species provisionally, *N. baileyi*, *N. caudatus*, and *N. goral*.

Introduction

The serows and gorals are probably the most primitive living caprine bovids. Their classification has been summarised most recently by Haltenorth (1963) and Dolan (1963). We here present conclusions from our own revisionary studies, based on the examination of skins and skulls in American and European museums.

Generic Status

Serows and gorals are usually assigned to separate genera, *Capricornis* Ogilby, 1837, and *Nemorhaedus* (*Naemorhedus* Hamilton Smith, 1827 — but this is a misspelling). Gorals and serows differ consistently in some characters, particularly features of the skull (Pocock, 1910). These differences are somewhat obscured by the contrasting proportions of the small and large-sized serows. Some of the apparent resemblances between the small serows and gorals (Pocock, 1908) may be regarded as superficial, for they relate to similarity in size.

Compared with serows, gorals have the facial part of the skull more flexed relative to the cranial part. The crista facialis (upper edge of masseter muscle origin) is more marked and extends further up on the face, pinching in the skull below the orbits, and ending in a prominent masseteric knob. The premaxillae are more robust and are parallel rather than converging distally. Most strikingly, the nasals are usually quite separate from the lacrimals and maxillae, producing a deep narial notch; the nasofrontal suture takes a different form; and the nasal branches of the premaxillae are much longer. These cranial differences probably reflect feeding specialisations. In addition, the lacrimal in gorals is narrowed dorsoventrally and shows little sign of a fossa, and the frontals form short, stout, tapering pedicles, so that the horn core bases are further from the orbital rim. All these differences appear to be derivative, and so are characteristics of the skin (reduced rhinarium, reduced preorbital glands, specialised interdigital glands).

Phylogenetic Problems

Gorals could have evolved from a species of serow, or the two groups of species could have evolved from a common ancestral taxon. The first hypothesis seems preferable, for some suggestive resemblances between the small serows and gorals remain when size similarities are taken into account. There is evidently a case for merging the genera *Capricornis* and *Nemorhaedus*, the latter name having priority.

Species of Nemorhaedus

N. *(*Capricornis*)* sumatraensis *(Figure 3.1)*

The serow is the largest species of the genus, with very long ears, long limbs, a prominent neck mane, very coarse pelage, and a long prognathous skull. Subspecies are *sumatraensis* (synonym

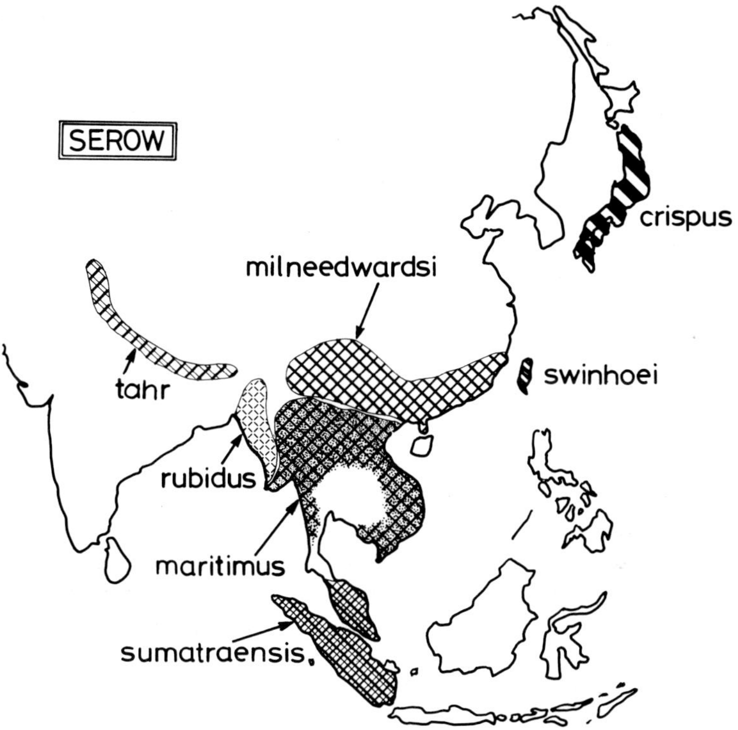

Figure 3.1: Distribution of the Serows *Nemorhaedus crispus, N. swinhoei* and *N. sumatraensis* with its Subspecies

swettenhami) of Malaya and Sumatra; *maritimus* (*annectens*), Indo-China, Thailand, and Burma; *milneedwardsi* (*argyrochaetes*, *montinus*), China; *rubidus*, southern Burma; and *thar* (*rodoni*, *humei*, *jamrachi*), Himalayas. Synonyms given in this paper are for subspecies which were still regarded as valid by Haltenorth (1963). Distribution and geographic variation are fairly continuous and the subspecies intergrade. They differ in size, colour of the mane, and intensity of pigmentation. (Figure 3.1)

N. *(C.)* swinhoei *(Figure 3.1)*

The Formosan serow, endemic to Taiwan, is one of the two smaller and monotypic species of serow. Its pelage is much softer than that of *N. sumatraensis*, it does not have a mane, and the ears are relatively shorter. A characteristic of this species is that even in relatively young adults, the upper edge of the lacrimal fossa usually forms a much more trenchant ridge than in other serows. The structure of the foot glands is not known.

N. *(C.)* crispus *(Figure 3.1)*

The Japanese serow of Honshu, Shikoku, and Kyushu is about the same size as *N. swinhoei*, but has much longer, softer hair, which forms a ruff on the neck, and still shorter ears. The narial notch is deeper than in other serows, with a longer nasal process of the premaxilla. The pedal glands are more specialised than in *N. sumatraensis*. These characters all suggest resemblances to the gorals. *N. crispus* is however unique in the large size of its cheek teeth.

N. *(*Nemorhaedus*) baileyi (Figure 3.2)*

The red goral is restricted to a small area of Burma, India, and China where these countries meet. The skin of the type specimen of *baileyi* is very similar to the specimens used for the description of another nominal species, *N. cranbrooki*, by Hayman (1961a) and was obtained just to the north of where *cranbrooki* has been observed. We believe that there is little doubt that the two taxa are conspecific. The skull of the type *baileyi* (Hayman, 1961b) resembles that of the other goral, except in the nasal region, where the nasal bones make a broad transverse suture with the frontals (as in serows), and are broadened proximally where they make extensive contact with the lacrimals. On one side, there is a trace of a suture separating this lateral flange of the nasal from the nasal bone proper, suggesting a separate centre of ossification in the formation of the bone. The condition appears to be intermediate between that of the serow *N. crispus* and the Chinese goral. Only a single skull assigned to *cranbrooki* has been described, and although fragmentary, the sutures appear to be more like the condition in other gorals (Hayman, 1961b).

Until the range of variation in skull characters of red gorals is known, two subspecies are recognised, *baileyi*, known so far only from Chinese Tibet, and *cranbrooki* of Assam and Upper Burma. *N. baileyi* separates other populations of gorals into a western Himalayan group, and an eastern Chinese group. No intergradation between these three is known and their morphological variation is greater than in the equally widespread *N.*

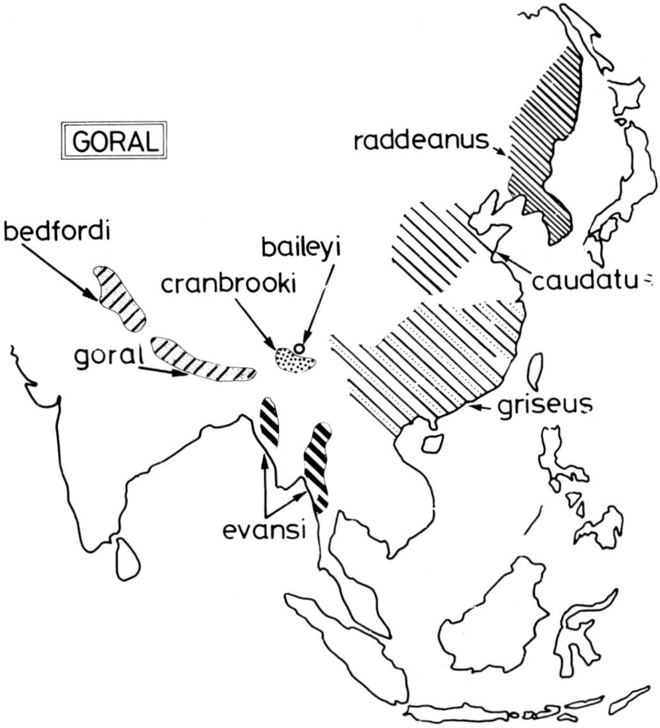

Figure 3.2: Distribution of the Gorals *Nemorhaedus caudatus*, with Subspecies *caudatus*, *griseus*, *raddeanus* and *evansi*; *N. baileyi*, with subspecies *baileyi* and *cranbrooki*; *N. goral*, with Subspecies *goral* and *bedfordi*

sumatraensis, so provisionally three species of goral are recognised.

N. *(*N.*)* caudatus *(Figure 3.2)*

This is the grey Chinese goral, which differs from the Himalayan in usually having longer pelage, especially in the race *raddeanus*, and a long tufted tail. The narial notch is often partly filled with wormian bones, while the frontal region of the skull is more inflated. Subspecies are *evansi*, a small race of Burma and Thailand, and *griseus* (*arnouxianus*), *caudatus*, and *raddeanus*, all of China, the last-mentioned extending into the Soviet Far East. Volf (1976) was unable to breed from *griseus* × *raddeanus* hybrids and so assumed that they were sterile and that the two kinds of goral are not conspecific. But *N. c. caudatus* as far as known is intermediate in pelage between the two, so Volf's conclusions need further investigation.

N. *(N.)* goral *(Figure 3.2)*

The Himalayan goral is if anything more specialised than *N. caudatus*. The skull is flatter, with a less convex facial region, the narial notch is deeper and never partly filled with wormian bones, and the masseteric ridge rises further up on the face, pinching in the skull more strongly. These characters are not absolute ones, but do apply to series of specimens. Two subspecies are recognisable, *bedfordi* ('*goral*') in the western Himalayas, and *goral* (*hodgsoni*) in the east.

References

Dolan, J.M. (1963) 'Beitrag zur systematischen Gliederung des Tribus Rupicaprini, Simpson 1945', *Zeit. Zool. Syst. Evolution*, *1*, 311–407

Haltenorth, T. (1963) 'Die Klassifikation der Säugetiere: Artiodactyla', *Handb. Zool., Berlin* (8), *32*, 1-167

Hayman, R. (1961a) 'The Red Goral of the North-east Frontier Region', *Proc. Zool. Soc., London*, *136*, 317-23

—— (1961b) 'Note on a Reputed Skull of *Nemorhaedus cranbrooki*', *J. Bombay Nat. Hist. Soc.*, *58*, 792-5

Pocock, R. I. (1908) 'On the Generic Names of the Rupicaprine Ruminants Known as Serows and Gorals', *Ann. Mag. Nat. Hist.*, *1*, 183-8

—— (1910) 'The Serows, Gorals and Takins of British India and the Straits Settlements', *J. Bombay Nat. Hist. Soc.*, *19*, 807-21

Volf, J. (1976) 'Some Remarks on the Taxonomy of the Genus *Nemorhaedus* H. Smith 1827 (Bovidae: Rupicaprini)', *Vert. Cesk. Spol. Zool.*, *40*, 75-80

CHAPTER 4

Notes on the Taxonomy of the Tatra Chamois (*Rupicapra rupicapra tatrica* Blahout)

V. HRABĚ and P. KOUBEK

Institute of Vertebrate Zoology, Czechoslovak Academy of Sciences, Květná 8, 60365 Brno (Czechoslovakia)

Abstract

Skulls and horns of 92 male and 40 female *R. r. tatrica* from the Tatra Mts (ČSSR) were analysed morphometrically to evaluate its subspecific status. Most characters showed sexual dimorphism. The shortcomings of the nasolacrimal fissure as a systematic character were demonstrated. For both sexes, most cranial dimensions of *R. r. tatrica* were significantly greater than those of the nominate subspecies, *R. r. rupicapra*.

Introduction

The chamois of the Tatra Mts, the only high mountains in Czechoslovakia, were described as a separate subspecies, *R. r. tatrica* by Blahout (1972). This is the northernmost autochthonous chamois population and the numbers of this mammal, rare in our fauna, have declined alarmingly in recent years (total population, 730 heads). Having examined more material than that available to Blahout, we obtained supplementary data and also evaluated males and females separately. Detailed results of this study have been published elsewhere (Hrabě & Koubek, 1984; Koubek & Hrabě, 1984). This is a short communication, reporting on the major results.

Material and Methods

We examined skulls of specimens older than 4 years (92 males, 40 females) of *R. r. tatrica* from the Tatra National Park. Most of the skulls came from specimens found dead over many years, a few from culled animals. They are kept in the collections of the Tatra National Park museum and of private persons. The mean individual age was 7.9 and 8.7 years for males and females, respectively. We analysed skull and horn dimensions as well as the shape and frequency of occurrence of the nasolacrimal fissure which is sometimes used as a subspecific character (Hrabě & Koubek, 1982, 1983).

Results

The means values of condylobasal length and that of the nasal part of the skull, horizontal diameter of orbit, biorbital and infraorbital width and neurocranium width and height as well as the mean length and girth of horns showed a marked sexual dimorphism. The nasolacrimal fissure (n.f.) was present in 70.4 per cent of males and 57.7 per cent of females. No correlation was found between the size and/or shape of the n.f. and individual age. The ratio of its mean length (9.3 mm) and mean width (4.7 mm) was 2 : 1. Both these dimensions were variously affected by the presence of apophysis lacrimalis which, in some cases, filled the space of the n.f. completely.

The mean values of most skull dimensions for *R. r. tatrica* (Table 4.1) were significantly higher than those of *R. r. rupicapra*, but the ranges of horn dimensions in the two subspecies overlapped in both sexes.

Discussion

Blahout's (1972) diagnosis of *R. r. tatrica* includes, among other characters, significantly greater means of total and basal length of skull, length and width of nasalia and length and width of n.f. as compared with *R. r. rupicapra*. As to horn dimensions, Blahout (1972) points out differences between *R. r. tatrica* and *R. r. rupicapra*.

Table 4.1: Basic Statistical Data on those Cranial Dimensions (mm) in which *Rupicapra rupicapra tatrica* Differs from *R. r. rupicapra* (data on *R. r. rupicapra* from Couturier, 1938; Sägesser, 1963-1965; Salzmann, 1977)

Dimension	Sex	n	min – max	$\bar{x}$	s	$s_{\bar{x}}$
Total length of skull	M	58	184.3-227.0	212.885	6.078	0.798
	F	33	204.0-222.0	211.133	4.519	0.787
Basal length	M	50	177.0-202.0	188.932	5.224	0.739
	F	32	175.0-198.0	187.350	5.231	0.925
Length of neurocranium base	M	11	45.0- 51.0	47.927	1.902	0.573
	F	6	47.0- 49.0	48.000	0.646	0.264
Biorbital width	M	78	98.7-117.4	110.086	4.488	0.508
	F	34	96.6-113.5	106.291	3.733	0.640
Infraorbital width	M	41	74.1- 84.0	79.185	3.293	0.514
	F	17	69.1- 81.1	76.124	3.104	0.753
Maxillary width	M	37	62.2- 70.0	66.765	1.965	0.323
	F	13	62.7- 70.1	66.892	2.485	0.689
Zygomatic width	M	61	74.3- 94.3	88.410	3.424	0.438
	F	32	73.0- 93.4	87.050	3.834	0.678

All means significantly greater than those for *R. r. rupicapra* at the $p<0.01$ level (Student's t - test).

Additional differences were shown by a comparison of our own data on *R. r. tatrica* with those on *R. r. rupicapra* (Table 4.1). The importance of these differences is emphasised by the fact that the series underlying Blahout's and our studies did not differ by sex or age structure. One of the causes of the differences in some skull dimensions is in their separate evaluation for males and females, which was not included in Blahout's study.

The horns of *R. r. tatrica* showed marked sexual dimorphism. No differences were revealed by a detailed comparison of horn dimensions and horn growth dynamics in *R. r. rupicapra* both autochthonous and introduced and *R. r. tatrica*. The horn dimensions of male and female *R. r. tatrica* overlap with the variation range of those in *R. r. rupicapra* from the populations compared, namely from Austria, Liechtenstein, Yugoslavia, Jeseníky Mts. in Czechoslovakia. Hence, we cannot recommend horn dimensions as a subspecific criterion in *R. r. tatrica* (Lomnicki, 1964; Blahout, 1972). Also the shape and size of the n.f., as characterised by Couturier (1938) and as currently used, cannot be used as a subspecific character according to our observations in *R. r. rupicapra* (both introduced and autochthonous), *R. r. tatrica* and *R. r. caucasica*. The example of *R. r. tatrica* shows how differently this character can be evaluated. Blahout (1972) found the presence of an n.f. in all skulls examined, taking even those filled with the lacrimal apophysis for positive. In our series the frequency of occurrence of the n.f. was considerably lower. We even found different dimensions of the n.f. so that the length: width ratio in our series was 2 : 1 in contrast to Blahout's 3.2 : 1.

Because of its considerable variability of occurrence, we consider the nasolacrimal fissure as unsuitable for the separation of *R. r. tatrica* from *R. r. rupicapra*, even though this character was used, e.g. in the descriptions of *R. r. caucasica* and *R. r. asiatica*.

An unequivocal determination of the subspecific status of the Tatra Mts chamois population requires a detailed examination of *R. r. carpatica* as well. Unfortunately, the knowledge of this subspecies is based solely on the very scanty material studied by Couturier (1938).

In conclusion, although having obtained somewhat different results from those used by Blahout (1972), we agree with his opinion that the Tatra Mts chamois population differs from the nominate subspecies in certain characters at the subspecific level (different coloration, certain somatic measurements and different ecological requirements, as well as in significantly greater skull dimensions) (Table 4.1).

References

Blahout, M. (1972) 'Zur Taxonomie der Population von *Rupicapra rupicapra* (Linné, 1758) in der Hohen Tatra', *Zool. Listy*, *21*, 115-32

Couturier, M. (1938) *Le Chamois*, Arthaud, Grenoble

Hrabě, V. & Koubek, P. (1982) 'Craniometric Characteristics of the Introduced Population of Chamois *Rupicapra rupicapra* in the Jeseníky Mts', *Folia Zool.*, *31*, 227-41

—— (1983) 'Horns of the Chamois, *Rupicapra rupicapra rupicapra*, from the Jeseníky Mountains, Czechoslovakia', *Folia Zool.*, *32*, 23-31

—— (1984) 'Craniometrical Characteristics of *Rupicapra rupicapra tatrica* Mamm., Bovidae', *Folia Zool.*, *33*, 73-84

Koubek, P. & Hrabě, V. (1984) 'Morphometrical Characteristics and Horn Growth Dynamics in *Rupicapra rupicapra tatrica* Mamm., Bovidae', *Folia Zool.*, *33*, 289-312

Lomnicki, A. (1964) 'An Analysis of the Horns of the Chamois, *Rupicapra rupicapra* Linnaeus, 1758 from the Tatra Mountains', *Ochrona Przyrody*, *50*, 141-56

Salzmann, H.C. (1977) 'Gewicht, Schädelgrösse und Gehörnwachstum von Gemsen aus dem Jura und ein Vergleich mit anderen Populationen', *Z. Jagdwiss.*, *23*, 69-80

Sägesser, H. (1963-1965) 'Über die Schädelproportionen einer Gemspopulation aus dem Augstmatthorngebiet', *Jahrb. d. Naturhist. Mus. Bern*, pp. 204-21

CHAPTER 5

Revision of *Rupicapra* Genus. III. Electrophoretic Studies Demonstrating Species Distinction of Chamois Populations of the Alps from those of the Apennines and Pyrenees

G. NASCETTI,[1] S. LOVARI,[2] P. LANFRANCHI,[3] C. BERDUCOU,[4] S. MATTIUCCI,[5] L. ROSSI[3] and L. BULLINI[1]

1. Istituto di Genetica, Università di Roma, 00185 Roma (Italy)
2. Istituto di Zoologia, Università di Parma, 43100 Parma (Italy)
3. Istituto di Patologia Speciale e Clinica Medica Veterinaria, Università di Torino, 10126 Torino (Italy)
4. Centre d'Etude et Recherche sur la Faune de Montagne, 64880 Etsaut (France)
5. Istituto di Parassitologia, Università di Roma, 00185 Roma (Italy)

Abstract

The genetic structure of chamois populations from the Alps (*Rupicapra rupicapra rupicapra* L. 1758), Apennines (*R.r. ornata* Neumann 1899), and Pyrenees (*R.r. pyrenaica* Bonaparte 1845) was investigated by means of the electrophoretic analysis of 25 enzyme loci. Two diagnostic loci, *Me-1* and *Me-2*, were found distinguishing *rupicapra* from *pyrenaica* and *ornata*, each allowing the separation of these two groups with a probability of at least 99 per cent. Within the *pyrenaica-ornata* group, the genetic structure in *pyrenaica* is similar to that of *ornata*. The genetic divergence of the three taxa and the variability within each are analysed. The very low variability of *ornata* of the Apennines is likely due to a recent bottleneck of this endangered taxon; the moderately low variability of the abundant Alpine *rupicapra* is surprising. The presence of fully effective anti-hybridisation mechanisms between *pyrenaica-ornata* and *rupicapra* is suggested, on the basis of genetic, behavioural, and palaeozoogeographic evidence. Accordingly, two distinct species are indicated: *R. rupicapra* and *R. pyrenaica* (the latter including the subspecies *ornata*). An experimental study of their reproductive isolation in artificial sympatric areas in the field is proposed, supported by the confident genetic placement of resulting offspring by the use of biochemical markers.

Introduction

The genus *Rupicapra*, probably of Asiatic origin, differentiated in Europe and the Near East during the Pleistocene. The oldest fossil chamois belong to the late Mindel or the early Riss glaciations, from the eastern Pyrenees: Caune de l'Arago, but records become common only from Würm 2 onwards (cf. Masini, 1983). Ten subspecies of chamois are generally recognised, scattered on mountains of South-central Europe and the Near East (e.g. Knaus & Schröder, 1975). Some morphometric and morphological evidence indicates that the south-western chamois group together, separately from all other subspecies (Camerano, 1914; Lovari & Scala, 1980; Lovari, 1984). Differences in reproductive behaviour patterns between *R. rupicapra rupicapra* of the Alps and *ornata* of the Apennines, apparently not related to environmental features, have been outlined by Lovari (1981). At least one of *ornata*'s behaviour patterns (Dynamic Head Down) has been observed in *pyrenaica* as well (Lovari, unpublished observations), correlating with other evidence for a *pyrenaica-ornata* type at the morphological level: yellowish and dark brown winter coat pattern; small skull; close horn bases; obliteration of the fronto-maxillary fontanelle (Lovari & Scala, 1980). According to these authors, during the Würm glaciation the *pyrenaica-ornata* type may have come into contact with populations of the 'Alpine' type (whitish and very dark winter coat pattern; large skull; relatively divergent horn bases; usually open fronto-maxillary fontanelle), with possible geographically limited introgression phenomena (Lovari & Scala, 1980). Several differences at the morphometrical level between *rupicapra* and *ornata* populations have been described by Fondi *et al.* (in preparation), while significant differentiation in the horns and skull between *ornata* and *pyrenaica* has been recently demonstrated by Scala & Lovari (1984) at subspecific level.

The aim of this paper is to show the genetic relationships among natural populations of *rupicapra*, *pyrenaica* and *ornata* by means of multilocus electrophoretic analysis. This approach has provided data of critical importance to the evolutionary biology and taxonomy of several animal and plant groups, as shown in recent papers (Coluzzi & Bullini, 1971; Avise, 1974; Ayala, 1975; Bullini & Sbordoni, 1980; Ferguson, 1980; Wake, 1981; Bullini, in press).

The authors express their gratitude to F. Tassi for making possible the collection of samples in the Abruzzo National Park; to T. Balbo and E. Biocca for their helpful suggestions and stimulating encouragement; to R. Cianchi for her critical reading of the manuscript; to R. Fico, A. Rossetto, G. Tosi, the direction and personnel of the Belviso e Barbellino. Valbondione e Malgina, Monte Bianco hunting reserves as well as the Gruppo Faunistico del C.F.S. Foresta Demaniale di Tarvisio, for their help in the collection of samples. S.L. was supported by a grant M.P.I.

Material and Methods

The *Rupicapra* populations studied and their locations are as follows: *rupicapra*, Eastern Alps (sample size, N = 7), Central Alps (N = 12), Western Alps (N = 24); *pyrenaica*, Central Pyrenees (N = 25); *ornata*, Central Apennines (N = 18).

Horizontal starch gel electrophoresis was performed on muscle tissue, obtained either from chamois shot by hunters (most Alpine and Pyrenean specimens), or by surgically removing about 1 g of muscle from the rump of darted and sedated individuals.

The following enzymes were assayed, for a total of 25 loci (indicated in brackets): α-glycerophosphate dehydrogenase (*α-Gpdh*), sorbitol dehydrogenase (*Sdh*), lactate dehydrogenase (*Ldh-1*, *Ldh-2*), malate dehydrogenase (*Mdh-1*, *Mdh-2*), malic enzyme (*Me-1*, *Me-2*), isocitrate dehydrogenase (*Idh-1*, *Idh-2*), 6-phosphogluconate dehydrogenase (*6Pgdh*), glyceraldehyde-3-phosphate dehydrogenase (*G3pdh*), indophenol oxidase (*Ipo-1*, *Ipo-2*), glutamate-oxaloacetate transaminase (*Got-1*, *Got-2*), creatine kinase (*Ck*), adenylate kinase (*Adk*), phosphoglucomutase (*Pgm-2*), esterase (*Est-1*, *Est-2*, *Est-3*), adenosine deaminase (*Ada*), mannose phosphate isomerase (*Mpi*) and phosphoglucose isomerase (*Pgi*). The techniques used were, with minor modifications, those described by Brewer (1970), Selander *et al.* (1971), Ayala *et al.* (1972), and Harris & Hopkinson (1976).

Three parameters: expected mean heterozygosity (H), proportion of polymorphic loci, 1 per cent criterion (P), and mean number of alleles per locus (A) were used to quantify the genetic variability of populations. The degree of genetic divergence between taxa was estimated with Nei's indices of standard genetic identity: *I*, and standard genetic distance: *D* (Nei, 1972).

Results and Discussion

18 out of the 25 loci have been found monomorphic for the same allele in all the samples tested; allele frequencies at the remaining loci: *Me-1*, *Me-2*, *Got-1*, *Est-1*, *Est-2*, *Est-3*, and *Mpi* are given in Table 5.1. *Me-1* and *Me-2* appear to be diagnostic between *rupicapra* and *pyrenaica-ornata*, each allowing a correct identification with a probability of at least 0.99. A third locus, *Est-1*, shows significantly different allele frequencies between the two groups. Within the latter group, the genetic structure in *pyrenaica* has been found to be similar to that of *ornata*. The values of Nei's genetic distance (*D*) and identity (*I*) between the three taxa are given in Table 5.2. *D* ranges from 0.009 (between *pyrenaica* and *ornata*) to 0.121 (between *ornata* and *rupicapra*). Populations belonging to *R. rupicapra rupicapra* appear genetically similar, in spite of their geographic spread (e.g. $D = 0.001$ between the Eastern and Western populations, at the two ends of the Alps); this appears to be related to a panmictic condition of the Alps chamois during the last glaciation.

Different degrees of genetic variability have been found in chamois from the Alps, Apennines, and Pyrenees (Table 5.3). The most variable taxon appears to be *pyrenaica*, with $H = 0.033$, $P = 0.20$, and $A = 1.20$; similar values have been observed in other mammals (e.g. Powell, 1975; Nevo, 1978). The absence of genetic variation found in *ornata* may be accounted for by its small population size (350-400 individuals) and the short time since the last bottleneck that, 35 years ago, reduced this taxon to about 40 individuals (Lovari, 1977). The intermediate variability observed in *rupicapra* from the Alps has no obvious explanation; population size is large there: more than 340,000 (while being about 15,000 in the Pyrenees) (Lovari, 1984), and no evidence of recent bottleneck events exists for this taxon.

The electrophoretic data show that a considerable genetic divergence exists between *rupicapra* and the *pyrenaica-ornata* group: $D = 0.108$, a value observed both between well-differentiated subspecies and distinct species of many groups of animals. A much lower genetic divergence has been found, on the contrary, between *pyrenaica* and *ornata* ($D = 0.009$), indicating that gene flow took place between populations of these two taxa until recently, presumably in the Würm glaciation. Apparently no significant gene flow occurred at that time

Table 5.1: Allele Frequencies at Seven Enzyme Loci in Populations of Chamois of Different Geographic Origin. REA = Eastern Alps; RCA = Central Alps; RWA = Western Alps; OAB = Central Apennines; PCP = Central Pyrenees

Loci	Alleles	REA	*rupicapra* RCA	RWA	*ornata* OAB	*pyrenaica* PCP
Me-1	*93*	—	—	—	1.00	1.00
	100	1.00	1.00	1.00	—	—
Me-2	*100*	0.93	1.00	0.92	—	—
	108	0.07	—	0.08	1.00	1.00
Got-1	*100*	1.00	1.00	1.00	1.00	0.98
	112	—	—	—	—	0.02
Est-1	*95*	—	—	0.13	1.00	0.58
	100	1.00	1.00	0.87	—	0.42
Est-2	*96*	—	—	—	—	0.06
	98	—	0.08	—	—	—
	100	1.00	0.92	1.00	1.00	0.94
Est-3	*93*	—	—	—	—	0.02
	100	1.00	1.00	1.00	1.00	0.98
Mpi	*100*	1.00	1.00	1.00	1.00	0.90
	106	—	—	—	—	0.10
	*H**	*0.005*	*0.006*	*0.015*	*0.000*	*0.033*

**H* = mean heterozygosity overall for 25 loci studied, which include 18 monomorphic loci not listed in the table (see text).

Table 5.2: Average Values of Nei's (1972) Genetic Identity (above the Diagonal) and Genetic Distance (below the Diagonal) between *rupicapra*, *pyrenaica* and *ornata*

	R.r. rupicapra	*R.p. pyrenaica*	*R.p. ornata*
R. rupicapra rupicapra	—	0.909	0.886
R. pyrenaica pyrenaica	0.096	—	0.991
R. pyrenaica ornata	0.121	0.009	—

Table 5.3: Parameters of Overall Genetic Variability in Three *Rupicapra* Taxa, on the Basis of 25 Enzyme Loci. H = expected mean heterozygosity; P = proportion of polymorphic loci; A = mean number of alleles per locus

	H	P	A
R. rupicapra rupicapra	0.012	0.12	1.12
R. pyrenaica pyrenaica	0.033	0.20	1.20
R. pyrenaica ornata	0.000	0.00	1.00

between *rupicapra* and the *pyrenaica-ornata* group, although it is likely that there were sympatric areas (Lovari & Scala, 1980; Masini, 1983). This suggests that anti-hybridisation mechanisms between *rupicapra* and *pyrenaica-ornata* were operating. Accordingly, it is apparent that two distinct species: *R. rupicapra* and *R. pyrenaica* (the latter including the subspecies *ornata*) exist. The study of their reproductive isolation will require field release experiments which make sympatric the two supposed species in confined biotopes. The diagnostic loci (*Me-1* and *Me-2*) found between *rupicapra* and *pyrenaica-ornata* allow the positive analysis of possible hybridisation and/or introgression phenomena, and will make easier the interpretation of such an experiment. A similar approach has been successfully used to test the reproductive isolation between mosquitoes of the *Aedes mariae* complex (Coluzzi & Bullini, 1971).

References

Avise, J. (1974) 'Systematic Value of Electrophoretic Data', *Syst. Zool.*, *23*, 465-81

Ayala, F.J. (1975) 'Genetic Differentiation during the Speciation Process', *Evol. Biol.*, *8*, 1-78

——, Powell, J.R., Tracey, M.L., Mourão, C.A. & Pérez Salas, S. (1972) 'Enzyme Variability in the *Drosophila willistoni* Group. IV. Genic Variation in Natural Populations of *Drosophila willistoni*', *Genetics*, *70*, 113-39

Brewer, G.I. (1970) *An Introduction to Isozyme Techniques*. Academic Press, London

Bullini, L. (in press). 'Taxonomic and Evolutionary Inferences from Electrophoretic Studies of Various Animal Groups' in D. Rollinson & G.S. Oxford (eds.) *Adaptation and Taxonomic Significance of Protein Variation*, Academic Press, London

Bullini, L. & Sbordoni, V. (1980) 'Electrophoretic Studies of Gene Enzyme Systems: Microevolutionary Processes and Phylogenetic Interference', *Boll. Zool.*, *47* (suppl.), 95-112

Camerano, L. (1914) 'Ricerche Intorno ai Camosci', *Mem. R. Accad. Sci. Torino. (Cl. Sci. Fis. Mat. e Nat., s. II)*, *65*, 1-82

Coluzzi, M. & Bullini, L. (1971) 'Enzyme Variants as Markers in the Study of Precopulatory Isolating Mechanisms', *Nature*, 231, 455-6

Ferguson, A. (1980) *Biochemical Systematics and Evolution*, Blackie & Son, Glasgow

Harris, H. & Hopkinson, D.A. (1976) *Handbook of Enzyme Electrophoresis in Human Genetics*, North Holland Publ., Amsterdam

Knaus, W. & Schröder, W. (1975) *Das Gamswild*. Paul Parey, Hamburg

Lovari, S. (1977) 'The Abruzzo Chamois', *Oryx*, *14*, 47-50

—— (1981) 'Behaviour as an Aid to Systematics. The Case of the Abruzzian Chamois *Rupicapra* (*rupicapra*) *ornata* Neumann 1899', *Abst. XVIIth Ethological Conf.*, Oxford

—— (1984) *Il Popolo delle Rocce*, Rizzoli, Milan

—— & Scala, C. (1980) 'Revision of *Rupicapra* Genus. I. A Statistical Re-evaluation of Couturier's Data on the Morphometry of Six Chamois Subspecies', *Boll. Zool.*, *47*, 113-24

Masini, F. (1983) *I Camosci Subfossili del Monte Sumbra (Alpi Apuane)*. Tesi di laurea in Scienze Biologiche, Univ. Firenze

Nei, M. (1972) 'Genetic Distance between Populations', *Amer. Natur.*, *106*, 283-92

Nevo, E. (1978) 'Genetic Variation in Natural Populations: Patterns and Theory', *Theor. Pop. Biol.*, *13*, 121-77

Powell, J.R. (1975) 'Protein Variation in Natural Populations of Animals', *Evol. Biol.*, *8*, 79-119

Scala, C. & Lovari, S. (1984) 'Revision of *Rupicapra* Genus. II. A Skull and Horn Statistical Comparison of *Rupicapra rupicapra ornata*, and *R.r. pyrenaica* chamois', *Boll. Zool.*, *51*, 43-52

Selander, R.K., Smith, M.H., Yang, S. Johnson, W.E. & Gentry, J.B. (1971) 'Biochemical Polymorphism and Systematics in the Genus *Peromyscus*. I. Variation in the Old Field Mouse (*Peromyscus polionotus*)', *Studies in Genetics, 6 Univ. Texas Publ.*, *7103*, 49-90

Wake, D.B. (1981) 'The Application of Allozyme Evidence to Problems in the Evolution of Morphology' in G.G.E. Scudder & J.L. Reveal (eds.) *Evolution Today, Proc. II Intern. Congr. Systematics and Evolutionary Biology*, 257-70

PART TWO

Ecology and Behaviour

CHAPTER 6

Correct Age Determination in Chamois

W. SCHRÖDER and IRMGARD VON ELSNER-SCHACK
Institute of Wildlife Research, University of Munich, Amalienstrasse 52, D-8000 München 40 (FRG)

The regularity of horn growth in chamois has been known for decades. These principles make it possible to determine the age of most animals. However the examination of the recorded ages for animals exhibited at many Trophy Shows in Austria and Bavaria revealed errors in the determination of age, as well as differing designation of individual age classes. The following contribution is intended to limit possible errors and to standardise the designation of age classes.

Horn Growth

A chamois horn grows every year, but only between spring and autumn. The separate growth rings are more or less sharply defined. It is usually the separation between the kid and yearling parts that is most difficult to recognise (Figure 6.1). The growth segments are related to one another in length. The yearling segment is the longest. Only rarely is it exceeded by the segment grown in the second year. Table 6.1 shows the length of the growth segments of chamois from the Ammergau Mountains.

The separate growth segments decrease in length rapidly after the age of one year. Only the two-, three- and four-year-old rings are relatively wide. Thereafter the segments are narrow and usually referred to as 'millimetre rings'.

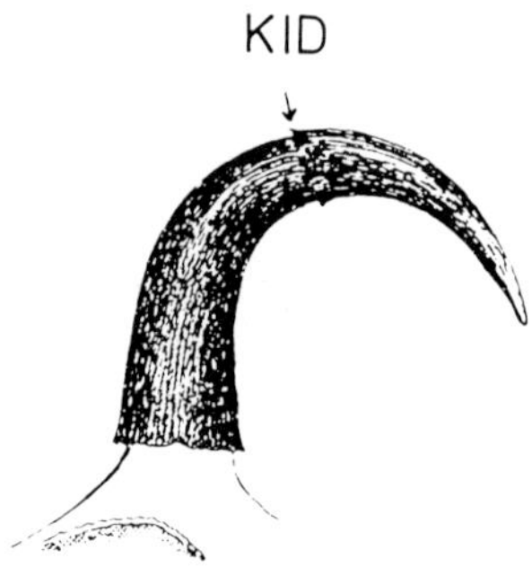

Figure 6.1: Horn of a Chamois Yearling with Two Growth Segments. The length of this animal's yearling segment is unusually long

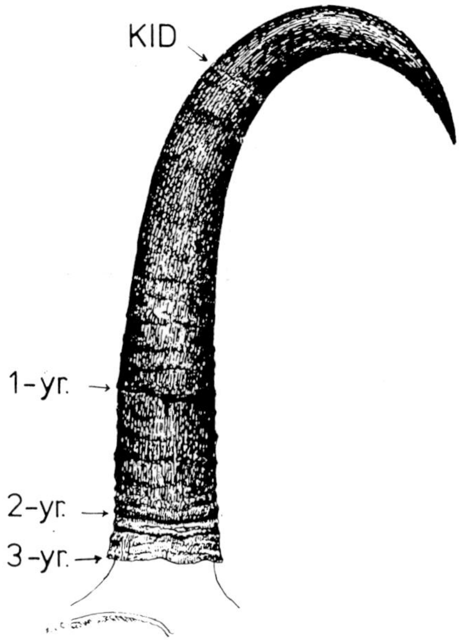

Figure 6.2: Three-year-old Chamois Male. The animal has two longer segments after that of the yearling. The ring-like swellings at the base of the horn (adornment rings, 'Schmuckringe'), therefore cannot be millimetre rings. Occasionally they are erroneously counted as such

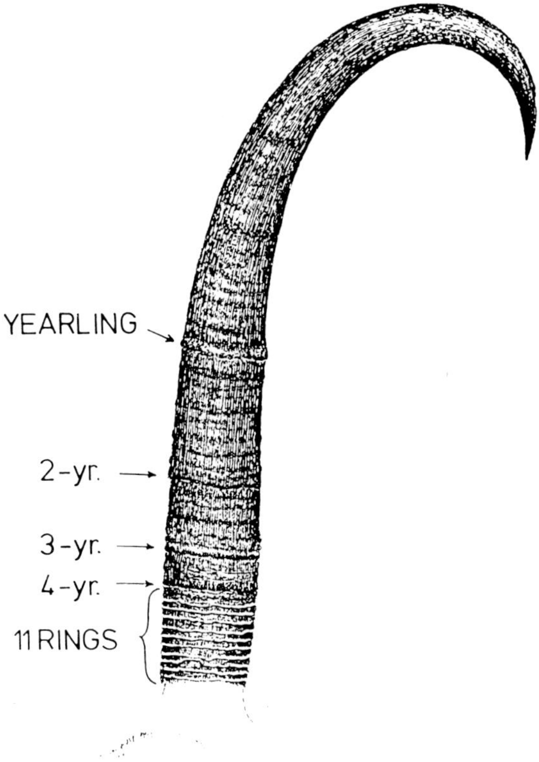

Figure 6.3: 15-year-old Chamois Male. The sequence of longer segments up to the age of four years has been marked with an arrow. Thereafter follow 11 'millimetre rings'

Table 6.1: Length of Growth Segments in Millimetres of 70 Chamois Horns

Age	Males Mean (Min.-Max.)	Females Mean (Min.-Max.)
Kids + yearlings	145 (69-177)	121 (99-150)
Two year old	40 (28-64)	37 (12-49)
Three year old	17 (5-30)	17 (8-31)
Four year old	5 (3-9)	5 (2-14)

Sources of Error

Most errors occur when it is not known that the segments form a decreasing sequence. Young males with a long yearling segment are often judged too old, usually because it is hard to believe that the animal is still so young. The male in Figure 6.2 is an example of this. In such cases, the adorning rings ('Schmuckringe') at the base of the horn are counted too, and so the age calculated is incorrect.

It is important to pay attention to the pattern of the relatively wide segments of the yearling, two, three and four year olds. One should not count the millimetre rings until these first four segments have been determined. In the case of an older animal, therefore, many experienced hunters first look for the conclusion of this sequence of the four wider rings. This means that the animal is four years old. Only then is the number of millimetre rings counted. The age of the animal is then 4 + the number of millimetre rings (Figure 6.3). To avoid making mistakes it is helpful to know exactly which ring one is looking for.

Standardised Designation

A chamois has one more growth segment than its age. The one-year-old animal has two: the kid segment and that of the yearling. The yearling is one full year old. If it is shot in November, it is exactly 1 year and 6 months old. To be correct only completed years of growth should be counted, as for roe deer, red deer, and, by the way, for humans too. Someone who is 'forty' has completed forty years of life.

In some areas chamois are habitually labelled one year too old; the yearling is referred to as a two year old.

At the Trophy Show in the county of Berchtesgaden, the reported age of chamois was compared with the actual age, as determined by experts. Figure 6.4 shows the result: of 37 males, 24 were reported one year too old. This is the systematic error in the designation of age. Eleven males were reported to be more than one year too old. These are errors in the determination of age. Only two males were correctly classified.

For females, of 38 animals 7 were classified correctly (Figure 6.5); 20 females were reported as one year too old. Here again,

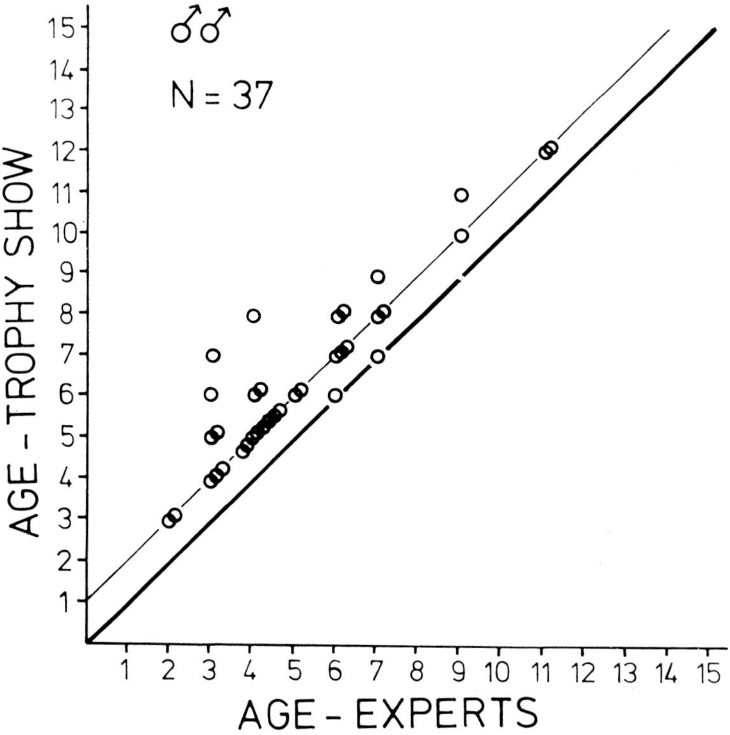

Figure 6.4: Comparison Between the Ages of Chamois Males Reported at the Trophy Show with those Determined by Experts. If both age estimations are the same, the points lie on the diagonal, thick line. This is true for only two animals. Most points lie one year off: these chamois were systematically reported one year older than they actually are (for example, the three-year-old male as four years old). Some males, usually three and four year olds, were reported several years too old

the erroneous designation of age is apparent. Six females were two years too old and five were too young.

Recommendation

In order to avoid errors in the determination of age, the illustrations presented here are intended to help the inexperienced who do not often have a chamois horn in hand. Once the

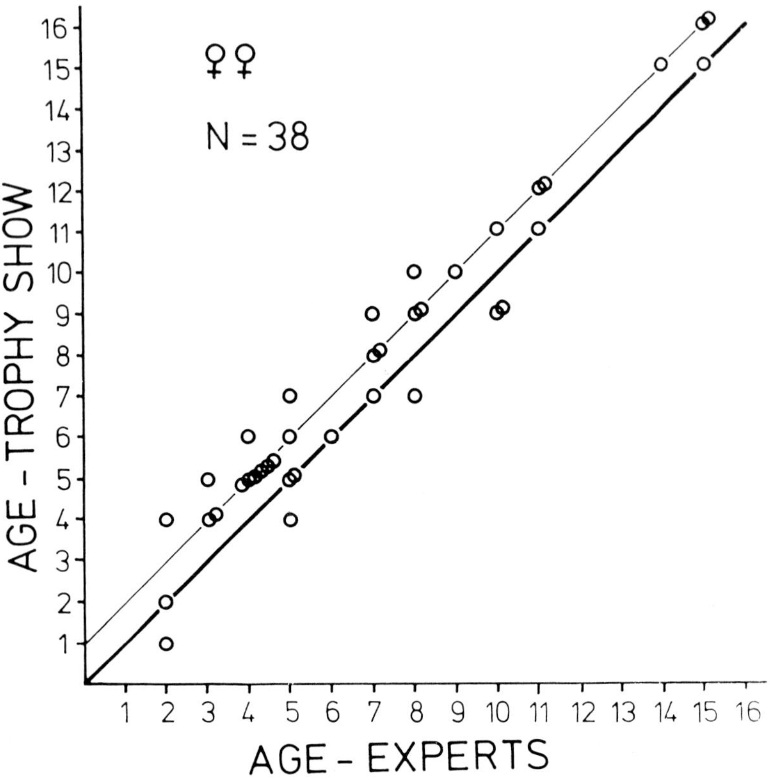

Figure 6.5: Comparison of the Ages of Females Reported at the Trophy Show with the Ages as Determined by Experts. Most females are reported one year too old. The reported ages of females are not so consistently too old as for males. Some are even reported too young

growth segments have been correctly identified, it is only a question of correct designation. Only completed years count: kids (0), yearlings (1), two-year-olds (2) . . .

The standardised designation of age for chamois is an important prerequisite for supra-regional comparisons of hunting harvests as well as for behavioural studies.

CHAPTER 7

What is Good Chamois Habitat?

IRMGARD VON ELSNER-SCHACK
Institute of Wildlife Research, University of Munich, Amalienstrasse 52, D-8000 München 40 (FRG)

Abstract

Habitat preferences were studied in a chamois population in a Bavarian mountain range. Animals exhibit seasonal patterns of altitudinal distribution, preferring higher altitudes in summer. Open vegetation forms were selected through all seasons, even in winter. A variety of slopes were utilised, eastern and southeastern directions most heavily. The extremes were selected, very steep and very flat, varying with snow cover. Food and security aspects determine habitat quality.

Introduction

We know where to find chamois, but we lack a working definition of chamois habitat. Habitat selection was studied over an annual cycle in the Ammergauer Berge. Habitat variables that were preferred at different seasons could thus be isolated and identified.

Study Area and Methods

The study area ranges from 900 m in the valley bottoms to 1890 m at the summits. The lower slopes are mostly forested: mixed mountain forest (*Picea abies*, *Abies alba*, *Fagus sylvatica*), coniferous forest (spruce and fir). Higher up there are open pine forests (*Pinus sylvestris*) and patches of knee pine (*Pinus mugo*), especially around timberline. The study area contains some extensive alpine pastures and natural alpine grasslands located in cirques and steep slopes. Because of geological conditions (dolomite) some of the area is very rugged with rocks and gravel.

Locations of chamois were recorded regularly as the season progressed and analysed afterwards to identify important habitat variables. This paper focuses on those habitat variables, such as altitude, vegetation, exposition, slope and rock that were used more frequently than could be statistically expected. (The null hypothesis assumes that chamois are distributed proportionally to available habitat types.) For details of altitudinal zoning and analysis see von Elsner (1980, 1982).

Results

Distribution of chamois shows a seasonal pattern with respect to altitude. During the winter months chamois were recorded above expectation about halfway up the mountains. During snow-free periods they clearly preferred the highest altitude around or above timberline (Figure 7.1, black outline).

Use of vegetation shows the following pattern: open vegetation types were utilised during snow-free periods in areas where chamois concentrated. Alpine grassland, pastures and gravel describe this habitat. During snow-covered periods there is still a tendency to select open vegetation types, but forests become more important. Coniferous forest with open canopy was the forest type most frequently utilised. In those altitudes where chamois were observed less frequently than expected, mixed mountain forest was often preferred (Figure 7.1).

Chamois were usually found to inhabit several expositions at any time. East and southeastern directions were favoured most frequently, north and northwest expositions rarely (Figure 7.2).

Steep slopes were clearly preferred when snow was on the ground. During snow-free periods the flat areas were selected (Figure 7.3). In one instance, in autumn, chamois were also found to select very steep and very flat terrain at the same time. During summer, the few animals at low altitudes were found to like steeper slopes than in their preferred altitudes. Data for the

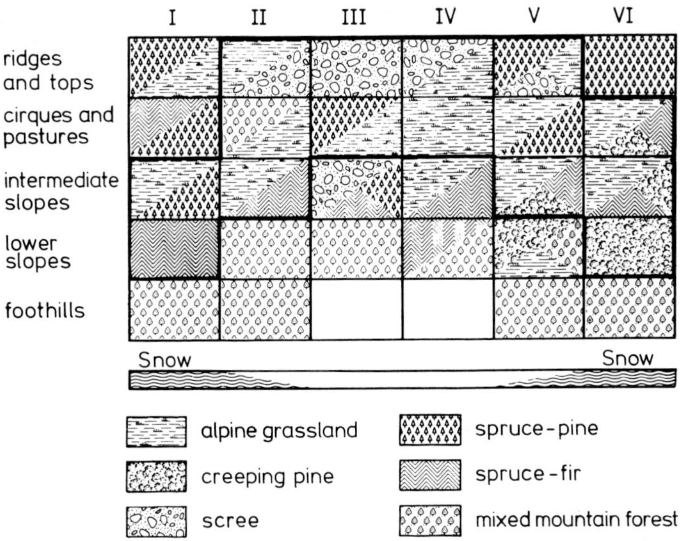

Figure 7.1: Preferred Vegetation Types According to Altitude and Season. In each altitutional zone up to three vegetation types were found to be more frequently preferred than expected. The heavy black line delineates preferred altitudes at different seasons. I = spring, II = early summer, III = late summer, IV = fall, V = early winter, VI = late winter

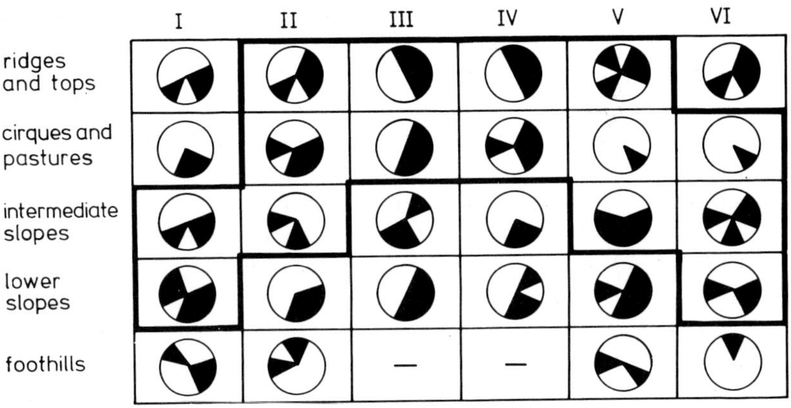

Figure 7.2: Preferred Expositions According to Altitude and Season. The heavy line delineates preferred altitudes. I – VI as in Figure 7.1

	I	II	III	IV	V	VI
ridges and tops	3	1	1	5/1	-	5/4/3
cirques and pastures	2/3	2	1	1	1/2	4/3
intermediate slopes	2/3	-	3	1/3	3	3
lower slopes	4/3	2	4/1	3	5/2/3	3
foothills	3/2	2	-	-	1/2	2

Figure 7.3: Preferred Slopes According to Altitude and Season. 1 = 1-20%; 2 = 21-40%; 3 = 41-60%; 4 = 61-80%; 5 = 81-100%. The heavy black line delineates preferred altitudes. I – VI as in Figure 7.1

affinity to rocks are not presented, because they correlate highly with slope.

Discussion

Several factors influence habitat selection. They can be classified as environmental constants, biotic and abiotic environmental variables and internal variables (Figure 7.4). The vegetation type, affecting the source of food and energy, seems to be of overriding importance. Open grasslands of various kinds satisfy this function. Aspects of security mean that the proximity of rocks is important. Because of security, rugged terrain interspersed with, or next to, grassland is also very important. This might be the reason why chamois do not use pastures in the valley bottoms, far away from rocks, although some pastures at an altitude of 900 m are well within the climatic tolerance of the species.

Because of seasonal pattern in vegetation development, chamois shift to vegetation of high quality. This is why they are found high up in the mountains during summer (see also Atzler, 1978) or select green vegetation on gravel beds when alpine meadows become brown towards the end of the growing season. When snow restricts access to otherwise preferred vegetation, chamois move to areas of steep, rocky slopes sometimes within

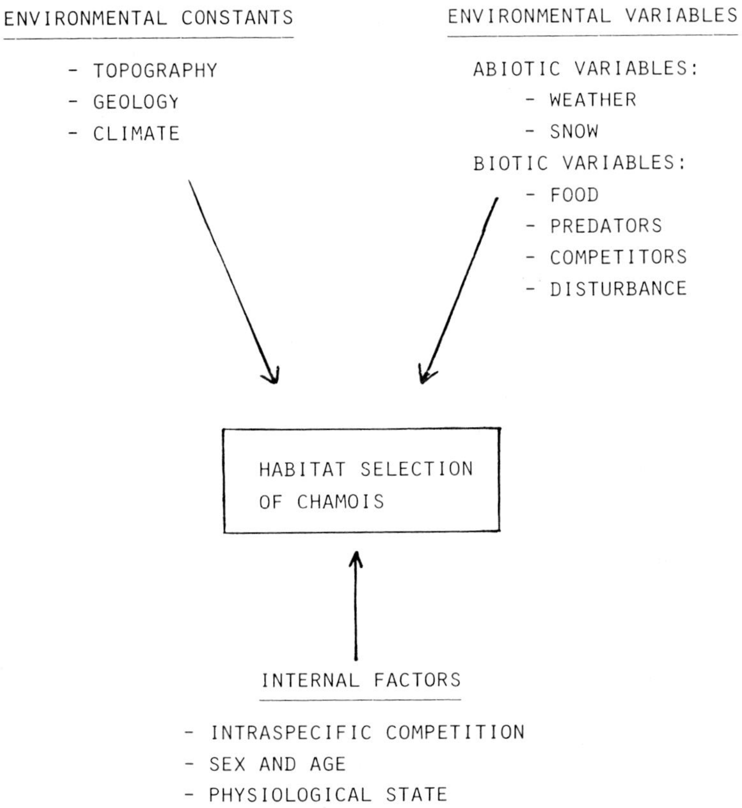

Figure 7.4: The Influences on Habitat Selection of Chamois

the forest. Understanding some of the requirements of the animal we may now characterise a chamois 'paradise': in summer it is an extensive alpine pasture with highly palatable plants and enough topographic variation to shift to new plant growth around patches of snow, shady places, gravel beds or northerly expositions. Sufficient rugged terrain is interspersed or close enough to provide security. Since feeding grounds are high up in the Alps, occasional wind will cool the animal during hot summer days and reduce insect harassment.

In winter, the ideal habitat includes steep, rocky places where little snow accumulates and with interspersed vegetation at expositions of high solar irradiation. Here animals survive in spite of deep snow elsewhere. These places can have some tree cover to protect from heavy winds and energy loss by radiation.

References

Atzler, R. (1978) 'Schwankungen der Qualität der Bodenvegetation während der Vegetationszeit am Rothirsch. Untersuchungsgebiet Ammergauer Berge', Dipl. Arbeit, Forstwiss. Fakultät, Universität München

Elsner von der Malsburg, I. (1980) 'Zur Raumnutzung von Gams (*Rupicapra rupicapra* L.). Untersuchungen an einer Population in den Ammergauer Bergen', Diss., Naturwiss. Fakultät, Universität Heidelberg

—— (1982) 'Überleben im Hochgebirge: Eine Studie zur Raumnutzung von Gams', *Z. f. Jagdwiss*, *28*, 18-30

Schröder, W. & Elsner, von der Malsburg, I. (1982) 'Gams (*Rupicapra rupicapra* L.) und Gebirgswald: Vorschläge zur Jagd auf Gams in den Ammergauer Bergen', *Forstwiss. Centralbl.*, *101*, 80-92

CHAPTER 8

Food Habits of Chamois (*Rupicapra rupicapra* L.) in Northern Tyrol

A. PERLE and J. HAMR
Foundation for Environmental Studies (FUST), 6215 Achenkirch (Austria)

Abstract

Food habits of a chamois population in an 8,000 ha study area were investigated from 1981 to 1983. Three methods were used: 1. rumen content analyses of shot animals; 2. direct observations of feeding animals; 3. control transect sampling. Chamois preferred grass and herbaceous plants to other vegetation during all seasons. During winters with deep, persistent snow, when ground vegetation was difficult to reach, chamois browsed on several tree species. Of these, the fir was most preferred, followed by maple, beech, pine and spruce respectively. The three methods of investigation yielded similar results as well as complementing one another.

Introduction

Chamois utilisation of forested alpine regions has been gaining more and more importance for forest managers in recent years. In some areas, browsing by ungulates on desired cultures of young trees has become a major forestry problem. It has been a general belief that the amount of damage done by chamois to young trees was insignificant.

This paper addresses the problem of chamois food habits at different times of the year. An attempt was made to bring to light the circumstances under which chamois invade valuable, young tree cultures for the purpose of browsing. The knowledge gained should offer some proposals for solutions to this forestry-wildlife management problem.

Study Area

The 8,000 ha study area is situated on the northern edge of the Tyrolean Limestone Alps. Altitudes range from 900 to 2,100 m. The region is characterised by high precipitation (average yearly precipitation varies between 1,450 and 1,850 mm). Continuous snow cover is present from the beginning of December until about mid-April. Average yearly temperature has been measured at +5.5°C in the valley (alt. 950 m). The tree line is located at altitudes of 1,500-1,600 m. Irregular patches of knee-pine (*Pinus mugo*) and mountain alder (*Alnus viridis*) grow up to the highest altitudes of the study area. Alpine grasslands, partially strongly overgrazed by cattle, as well as cliffs and fellfields are extensive above the tree line. The tree community is composed mainly of spruce (*Picea abies*), fir (*Abies alba*) and beech (*Fagus sylvatica*). Small pure stands of larch (*Larix decidua*) are common just below the tree line. Sparsely interspersed among the more numerous tree species are maple (*Acer pseudoplatanus*), ash (*Sorbus aucuparia*) and pine (*Pinus sylvestris*). Forest ground vegetation is composed mainly of grasses and sedges as well as shrubs of the family Ericaceae.

Chamois population density in the study area was estimated at 10 animals per 100 ha, red deer (*Cervus elaphus*) density at 3.5 animals and roe deer (*Capreolus capreolus*) density at 2.5 animals per 100 ha. The yearly mortality (natural and harvest) was estimated at 220 chamois, 110 red deer and 50 roe deer.

Funding was arranged by the Foundation for Environmental Studies (FUST — main contributor: Mr E. Underberg).

Materials and Methods

Three different methods of investigating the food habits of free-living chamois were used: (1) analyses of rumen contents of harvested animals; (2) field observations of feeding chamois; and (3) control transect sampling.

Rumen Contents

For the past year (1982-3), about ½ kg rumen contents were sampled from animals four times per month. Contents were placed onto a double sieve (upper layer, 3.15 × 3.15 mm mesh; lower layer, 1 × 1 mm mesh) and thoroughly washed with running water. Larger identifiable plant parts remained on the upper sieve, smaller finely chewed plant particles passed through onto the second. Contents of both sieves were dried at +80°C and weighed. The weight of each part as a percentage of the total was calculated. This method was recommended by the Institute for Wildlife Biology at the Veterinary School in Vienna. It differs from the method used by Kuen & Bubenik (1978).

Special attention was paid to the differences in food preferences of chamois shot above and below the tree line at different times of the year.

Field Observations

As a part of a more extensive study, observations of tagged and recognisable chamois were made for a period of two years. The animals were observed 1 to 6 times per week with a spotting scope (30×75) and binoculars (7×42). Choices of vegetation taken during feeding periods were recorded. Occasionally, chamois tracks were followed through the snow and notes made on the types of plants taken.

Transect Sampling

Ten permanently marked transects, 3 m wide and altogether 3,380 m long were periodically controlled for browsing damage done to young trees by red deer, roe deer and chamois during 1982. By observing the occurrence of animals on these transects at different times of the year and by following tracks and identifying faeces, it was possible, at least in some cases, to associate the browsing damage with a particular animal species. One of the transects in particular, where several males were observed on a limited area throughout the winter, served as a good example of chamois browsing preferences.

Observations from the transects were recorded on special forms supplied by the National Forestry Research Institute in Vienna. Among other information these forms included: descriptions of the topography, amounts and kinds of vegetation available, disturbance factors and the age structure of trees utilised in forestry practice. To facilitate comparisons between transects, the number of each tree species available and damaged per 1 ha was computed and presented graphically (Figures 8.1 and 8.2).

Results

Rumen Contents

This part of the study will be continued in the future. Preliminary results are shown in Table 8.1

Table 8.1: Preliminary Results of Rumen Content Analyses (1982 to 1983)

Plant species	Amount (%) Above the tree line summer	winter	in the forest summer	winter
Cyperaceae	14.6	31.5	6.7	55.0
Gramineae	31.3	4.6	51.4	5.3
Other herbaceous plants	17.0	7.3	39.2	2.3
unident. species of grass	33.3	11.9		
Polygala chamaebuxus	0.3	9.4	1.8	16.4
Vaccinium myrtillus	3.5	2.5		
Abies alba		1.6		13.8
Picea abies				6.0
Pinus sylvestris				0.3
Pinus mugo		9.9		
Fagus sylvatica				0.2
Sorbus aucuparia			0.9	
Rhododendron hirsutum		18.1		
Vaccinium vitis-idaea		2.4		
Juniperus communis		0.7		
Erica carnea		0.1		0.5
Lichenes				0.2

Clear preferences for grass and other herbaceous plants during the summer months are apparent. The only tree species found in the rumen contents during the period of vegetation abundance was the mountain ash. Preference for ground vegetation is still strong during the winter months when low shrubs of the family

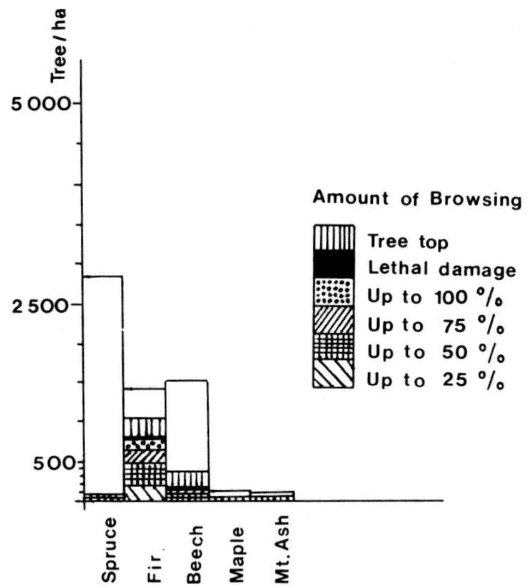

Figure 8.1: Distribution of Tree Species (white bars) and Browsing Intensity on a Portion of Transect Occasionally Frequented by Chamois

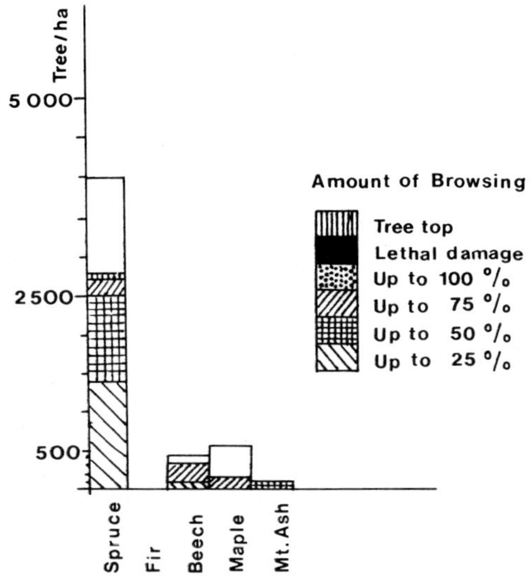

Figure 8.2: Distribution of Tree Species (white bars) and Browsing Intensity on a Portion of Transect Strongly Frequented by Chamois

Ericaceae also gain importance in the chamois diet. Especially striking in the winter is the role played by the shrub *Polygala chamaebuxus*. Of tree species, the one found most frequently in winter rumen samples was the fir, followed by the knee-pine and the spruce respectively.

Field Observations

Results gained by the rumen content analyses were largely confirmed by direct observations of feeding animals (Table 8.2). One inconsistency was presented by the lack of direct observations of browsing on the fir. This can be explained by the fact that in the region of the research area where most intensive observations of tagged chamois were made, young fir cultures were almost non-existent. The role of the spruce as a substitute for this lack was apparent. Tracking of chamois in the snow revealed relatively frequent use of low hanging twigs of mature beeches and of fallen pine branches, broken by the weight of snow or by wind. The described utilisation of beech and pine together with the scraping of lichens from the trunks of mature trees took place when a strong snow crust was present. Under these conditions, the ground vegetation under the snow was difficult to reach but motility on the snow surface was facilitated.

Transect Sampling

Figures 8.1 and 8.2 show browsing damage and availability for young trees on two portions of the same transect located inside a mostly male chamois winter range. The tree-culture represented

Table 8.2: Food habits of Chamois During One Year as a Percentage of each Month's Observations. 450 observations were made

Plant species	J	F	M	A	M	J	J	A	S	O	N	D
Grass	56	93	73	100	100	100	50	61	52	67	51	70
Herbaceous plants							33	31	38	33	26	
Picea abies	20		11						7			
Sorbus aucuparia							17	8				
Pinus mugo	4	4	2								13	10
Fagus sylvatica	12		4						1			10
Ericaceae											10	10
Lichenes, Bryophyta		3	5									
Pinus sylvestris	8		5									
Alnus viridis									2			

in Figure 8.1 was not particularly favoured by chamois because of its topography and snow conditions. Chamois concentrated frequently in the tree-culture represented by Figure 8.2 to feed and to rest. Figure 8.1 shows clear preference for the fir. About 75 per cent of the available young trees were browsed. The second most-preferred tree species was the mountain ash which is economically unimportant. Third was the weakly represented maple followed by the beech. The spruce, although well represented, was taken infrequently.

Figure 8.2 confirms the earlier assumption that the spruce becomes an important food item only when the preferred tree species are too few. In the culture represented by this diagram, the fir had been totally exterminated and the beech and the ash greatly reduced. Practically all the deciduous trees present were taken as soon as they appeared above the surface of the snow. Most of the maple represented in Figure 8.2 was very young and therefore permanently covered by snow during the winter. The transect surveys which yielded the results for Figures 8.1 and 8.2 were made during the snow-free period.

Discussion

None of the three methods used could alone answer the questions posed. Rumen content analyses give no information about the availability of plant species in a habitat. During field observations, species of plants cannot always be identified because of the large distances involved. In transect surveys, the animal species causing damage to trees cannot always be identified. Together, the three sets of results enhance and support one another.

Chamois rumen analyses done by Kuen & Bubenik (1978) in the same study area from 1970 to 1974 support for the most part the present results. These authors found comparatively few deciduous trees represented. This fact can be attributed to a radical increase of this group of trees in the study area between 1975 and 1980 due to altered forestry and wildlife management practices. Two further investigations of rumen contents, Onderscheka & Jordan (1974) in the eastern Alps of Austria, and Weyer (1974) in the Bavarian Alps, also produced results comparable to the present study.

In general, chamois do not browse on young trees if enough

grasses, sedges and low shrubs are in abundance and easily attainable. In the winter, when deep snow makes ground vegetation difficult to reach, chamois minimise energy expenditure by utilising readily reachable parts of trees above the snow surface.

References

Kuen, H. & Bubenik, A. B. (1978) 'Botanische Pansenanalysen bei Rotwild (*Cervus elaphus*), Rehwild (*Capreolus capreolus*) und Gamswild (*rupicapra rupicapra*)', *Beiträge zur Umweltgestaltung, 67*

Onderscheka, K. & Jordan, H. (1974) 'Einfluß der Jahrenszeit, des Biotops und der Äsungskonkurrenz auf die botanische Zusammensetzung des Panseninhaltes beim Gams-, Reh-, Muffel- und Rotwild', *1. Internationales Gasmwild — Treffen Oberammergau*, pp. 53-80

Weyer, U. (1974) 'Botanische Pansenanalysen bei Bayerischen Gemsen', *1. Internationales Gamswild — Treffen Oberammergau*, pp. 81-9

CHAPTER 9

Preliminary Observations on the Summer Diet of the Abruzzo Chamois (*Rupicapra rupicapra ornata* Neum.)

C. FERRARI and G. ROSSI
Istituto e Orto Botanico dell'Università, via Irnerio 42, I-40126 Bologna (Italy)

Abstract

The summer diet of the Abruzzo chamois (*Rupicapra rupicapra ornata* Neum.) has been studied in grassland areas above the timberline in the Val di Rose (Abruzzo National Park, Italy). The results were obtained by direct observations of grazing animals and by analysis of the plants browsed. The observations were made between June and September 1982. The results show a decreasing selectivity for both species and plant parts with the progress of summer into early autumn.

Introduction

This study shows some preliminary results obtained during a research programme on the feeding habits of the Abruzzo chamois (*Rupicapra rupicapra ornata* Neum.). Our results concern the diversity of plants grazed during the summer period in grasslands situated above the timberline within a specific study area, the upper Val di Rose, in the Abruzzo National Park (Italy), intensively frequented by this ungulate.

The major biological and ethological features of *R. rupicapra ornata* have been described in a number of papers (see Lovari, 1984, for a review). Several research activities are in progress by scientists coordinated by the 'Gruppo Camoscio Italia' under the sponsorship of the Abruzzo National Park.

Studies on the diet of the Abruzzo chamois are not available so far. Detailed information on the feeding behaviour of the Alpine chamois (e.g. Onderscheka, 1974; Onderscheka, Kutzer & Richter, 1968; Richard, 1972; Weyer, 1974; Dunant, 1977; Schröder, 1977) and of Pyrenean chamois (Berducou, 1975) reveal the usefulness of results which may clarify, in particular, the environmental impact of the species as well as the types of vegetation which can ensure the success of eventual reintroductions. In our case too, we hope to obtain this fundamental information as a result of our research.

We wish to thank F. Tassi and S. Lovari who have assisted our research in every possible way, as well as Lucia Naviglio, of the Centro Studi Ecologici Appenninici of the Abruzzo National Park, for her kind logistic assistance.

Study Area

The study area lies in the upper Val di Rose (Abruzzo National Park) between Monte Sterpi d'Alto (1,966 m) and Monte Boccanera (1,982 m). The high valley has the form of an extended basin with a predominant north-northwest exposure. This area is characterised by breaches formed by white and grey dolomitic limestone dating to the Lias (Praturlon, 1968). The present timberline of the beech forest is at approximately 1,700-1,800 m; on the rocks around the Passo Cavuto there are some isolated plants of *Pinus mugo*. The grasslands grazed by the Abruzzo chamois belong to the vegetation type described by Bazzichelli & Furnari (1970) as *Trifolium thalii* and *Plantago atrata* community, sometimes combining with the *Avena versicolor* and *Koeleria splendens* community.

Materials and Methods

Our observations were made during June, July and September 1982. Observations were not made in August on account of the tourists interfering with the behaviour of the herds.

The observed herds contained females and young chamois of both sexes. In September, the chamois born that year are also included. The sampling of the grazed species was carried out by direct observation. The herds were observed using 12×50 binoculars from a distance of about 30-40 m, taking note of various topographical references. For each of the grazed species, the parts of the plant were noted (F: flowers; L: leaves; W: whole plant), and the average grazing frequency is estimated (3: frequent to very frequent; 2: infrequent; 1: rare) in accordance with Dunant (1977).

Nomenclature of species follows Pignatti (1982).

Results

The sampled species are listed in Table 9.1. There is a total of 36 grazed species; and the total number of species is 73. This means that, during the observation period, the herds grazed 45.5 per cent of the total number of species.

The availability of a species is related to its phenological phase. For this reason, an analysis of grazing habits in each month is particularly revealing. Considering 'available' only those species which, in each month, were at a phenological phase unlike 'seedling' or 'dried plant', we see (Figure 9.1) that the number of available species decreases from June to September. At the same time the percentage of grazed species increases. This means that the herds are very selective during the month characterised by the greatest diversification of food (June). This behaviour is notably reduced in accordance with the progressive decrease of species in a suitable phenological phase. In September, as much as 83.8 per cent of the available species are grazed.

The data related to the parts of the plant eaten by the chamois are of particular interest (Table 9.1). We see that:

1. Only the flowers are eaten of the species *Doronicum columnae*, *Cirsium eriophorum*, *Carduus carlinaefolius*, *Senecio rupestris*, *Valeriana montana*, *Anthyllis vulneraria* s.l., *Armeria majellensis*, *Pulsatilla alpina* and *Rhinanthus alectorolophus*. When the flowering period comes to its end, these species are neglected, at least up to the end of September.
2. Only the leaves are grazed of the species *Plantago atrata*, *Veratrum album*, *Achillea millefolium*, *Verbascum longifolium*.

Table 9.1: Grazed and Ungrazed Species (Figures According to Dunant, 1977). Further explanations in the text

Grazed Species	June F	June L	June W	July F	July L	July W	September F	September L	September W	Ungrazed Species
Chenopodiaceae										Boraginaceae
Chenopodium bonus-henricus						2			2	*Myosotis alpestris*
Compositae										*Phyteuma orbiculare*
Doronicum columnae	2			3			1			Caryophyllaceae
Crepis aurea subsp. *glabrescens*		3				3			3	*Cerastium tomentosum*
Cirsium eriophorum	1			1			2			*Dianthus sylvestris*
Carduus carlinaefolius							1			Cistaceae
Senecio rupestris				1			1			*Helianthemum oelandicum* subsp. *alpestre*
Adenostyles australis				3					3	Compositae
Chrysanthemum tridactylites						3				*Leontodon hispidus*
Aster bellidiastrum						3				*Carduus chrysacanthus*
Achillea millefolium								2		*Carlina acaulis*
Cyperaceae										Campanulaceae
Carex kitaibeliana	1								1	*Campanula scheuchzeri*
Gentianaceae										*Bellis pusilla*
Gentiana lutea			1							Crassulaceae
Geraniaceae										*Sedum acre*
Geranium cinereum									3	Cruciferae
Geranium macrorrhizum									3	*Erysimum majellense*
										Gentianaceae
										Gentiana nivalis
Gramineae										*Gentiana verna*
Festuca robustifolia		2				1			1	Gramineae
Juncaceae										*Bromus erectus*
Juncus monanthos							1			*Phleum alpinum*
Labiatae										*Avena versicolor*
Stachys alopecurus subsp. *divulsa*										

Leguminosae									
Anthyllis vulneraria group	3		3						*Poa alpina*
Trifolium thalii					3			3	Juncaceae
Trifolium pratense subsp. *pratense*	1		2					2	*Luzula sieberi*
Trifolium repens					2			2	Labiatae
Liliaceae									*Stachys timphaea*
Veratrum album subsp. *lobelianum*							1		*Acinos alpinus*
Plantaginaceae									Leguminosae
Plantago atrata		1		3			2		*Medicago lupulina*
Plumbaginaceae									Ophioglossaceae
Armeria majellensis			1			1			*Botrychium lunaria*
Polygonaceae									Pinaceae
Rumex acetosa	3				3			1	*Juniperus nana*
Primulaceae									Polygalaceae
Soldanella alpina								1	*Polygala alpestris*
Ranunculaceae									Rosaceae
Ranunculus apenninus			2					2	*Geum urbanum*
Pulsatilla alpina subsp. *alpina*	1		1						*Alchemilla nitida*
Scrophulariaceae									*Alchemilla colorata*
Rhinanthus alectorolophus	1		1			1			Rubiaceae
Verbascum longifolium							2		*Galium anisophyllum*
Linaria purpurea						1			Santalaceae
Thymeleaceae									*Thesium parnassi*
Daphne mezereum					1				Scrophulariaceae
Umbelliferae									*Euphrasia salisburgensis*
Heracleum pyrenaicum subsp. *orsini*					3			3	*Scrophularia scopolii*
Pimpinella alpestris					2			2	*Veronica chamaedrys*
Valerianaceae									
Valeriana montana			3			3			
Violaceae									Umbelliferae
Viola eugeniae subsp. *eugeniae*					1				*Chaerophyllum hirsutum* subsp. *magellense*
									Urticaceae
									Urtica dioica

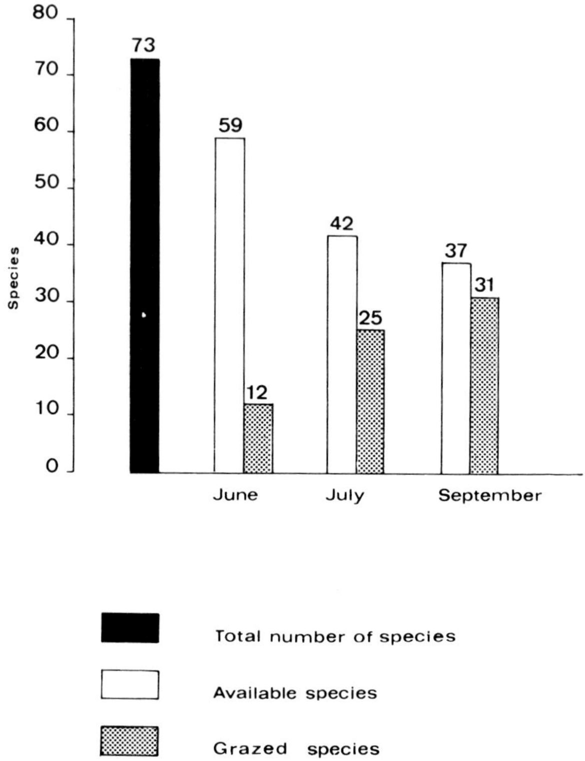

Figure 9.1: Number of Grazed Species During June, July and September in Relation to the Total Number of Species and the Available Species. (Further explanations in the text)

3. The species of which, initially, only the flowers or leaves are eaten and which in the course of summer are progressively subject to an indiscriminate form of grazing (the whole plant: W) are: *Crepis aurea subsp. glabrescens*, *Carex kitaibeliana*, *Festuca robustifolia*, *Trifolium pratense*, *Rumex acetosa* and *Ranunculus apenninus*.

We can synthesise as follows (Figure 9.2): the number of species of which only the flowers are eaten is constantly high in spite of a natural decrease in availability from June to September. During the period from June to July there is a rapid increase in the number of species of which all parts are grazed. A more moderate increase takes place from July to September.

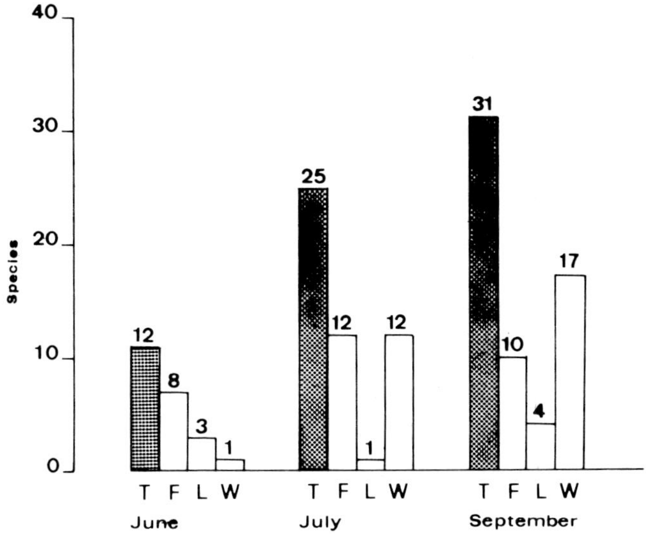

Figure 9.2: Number of Grazed Species (T) Subdivided According to Parts Eaten: flowers (F); leaves (L); whole plant (W). Further explanations in the text

Discussion

The results of our preliminary study give an indication of some significant tendencies in the feeding habits of the Abruzzo chamois from June to September.

According to other data (Locati & Lovari, unpublished data) we think that a few other species might be included in the summer diet of this chamois (e.g. *Hippocrepis comosa*, *Pedicularis comosa*). This fact, however, does not affect the main tendencies which are shown by our results.

The herds we have observed, comprising females and young animals of both sexes, are highly selective as long as this is possible. They become gradually less selective of plant species and parts eaten with the progress of summer into early autumn.

In general, we can say that, during the period of maximum species diversity and flowering activity (June), there is a clear preference for the flowers and, to a minor extent, for the leaves. When the number of available species declines, the chamois become less selective as to the parts preferred: the wholly grazed species amount is high in July and prevails in September. In particular, we can note that, among the grazed species, there is a high number of species of which only the flowers are eaten. During the less wet period (July) the number of species of which only the leaves are eaten is very low.

These tendencies clearly indicate that the phenology of the grasslands frequented by this ungulate has a significant role in determining the feeding preferences. This aspect of the problem will be carefully considered in our next studies (see also Dunant, 1977). Our observations, in particular, were made during the grazing period of highest food availability (period III according to Dunant, 1977) characterised by the flowering of dicotyledons and the widespread leaf growth of all species. Notwithstanding the relative shortness and the homogeneous features of this period, our monthly observations point out some significant shifts in the diet of the Abruzzo chamois.

References

Bazzichelli, G. & Furnari, F. (1970) *Ricerche sulla Flora e sulla Vegetazione di Altitudine nel Parco Nazionale d'Abruzzo*. Pubbl. Ist. Botanica Univ. Catania, Catania

Berducou, C. (1975) 'L'alimentation Hivernal de l'isard', *Bull. Off. Nat. Chasse*, Special, *4*, 195-227

Dunant, F. (1977) 'Le Régime Alimentaire du Chamois des Alpes (*Rupicapra rupicapra rupicapra* L.): Contribution Personnelle et Synthèse des Données Actuelles sur les Plantes Broutées', *Revue Suisse Zool.*, *84*, 883-903

Lovari, S. (1984) *Il Popolo delle Rocce*, Rizzoli, Milan

Onderscheka, K. (1974) 'Ernährungsprobleme bein Gamswild', *Tag. Ber. 1. int. Gamswildtreffen, Oberammergau*, 34-52

——, Kutzer, E. & Richter, H.E. (1968) 'Die Räude Gemse und ihre Bekämpfung; II Zusammenhänge zwischen Erhährung und Räude', *Z. Jagdwiss*, *14*, 12-27

Pignatti, S. (1982) *Flora d'Italia*, Edagricole, Bologna

Praturlon, A. (1968) *Note Illustrative della Carta Geologica d'Italia. Foglio 152 (Sora)*, Servizio Geologico d'Italia, Napoli

Richard, J.L. (1972) 'Un Facteur Écologique Nouveau dans le Jura; le Chamois', *Bull. Soc. Neuchatel. Sci. nat.*, *95*, 89-95

Schröder, W. (1977) 'Räumliche Verteilung und Nahrungsauswahl von Gams und Rotwild im Hochgebirge', *Forstw. Cbl.*, *96*, 94-9

Weyer, U. (1974) 'Botanische Pansenanalysen bei bayerischen Gemsen', *Tag. Ber. 1. int. Gamswildtreffen, Oberammergau*, 53-80

CHAPTER 10

Time Budgeting, Range Use Pattern and Relationships Within Groups of Individually Marked Chamois

T. PACHLATKO and B. NIEVERGELT
Ethologie & Wildforschung, Universität Zürich, Winterthurerstr. 190, 8057 Zürich (Switzerland)

Abstract

A female-young herd of chamois, including 27 individually marked animals, was observed in Aletschwald, canton Wallis, Switzerland, from July to October 1978. Seven females with kids were taken as focal animals using 'instantaneous sampling'. Two peak feeding periods were observed each day; these were more marked in September/October than in July/August. Feeding time was increased in September/October at the expense of 'standing and moving'. Focal female deviations from mean behaviour were ± 12 per cent for feeding and ± 8 per cent for lying. The occupied range varied between 21 and 28 ha each month. Two well-concealed rest-areas were used during the midday rest period. All relationships between marked animals were weak, except mother-kid bonds.

Introduction

Individually marked chamois are necessary to become acquainted with the social structure of this species. Marking animals costs a lot both in time and money, so it is not surprising that few data have been collected from individually distinguishable chamois (Krämer, 1969a).

This paper presents results from a diploma study by the first author and supervised by the second author (Pachlatko, 1980) concerning behavioural and ecological aspects in a herd of female and young chamois containing individually marked animals. Deviations by individual chamois from the mean 'normal behaviour' were a main subject of this study. Mean activity patterns of all animals in the herd, and deviations of some animals from this mean, are presented. Shifting of the home range between July and October is shown and relationships between marked chamois will be discussed. We would like to thank Mr R. Ittig and Mr H. Blatter for co-operation in the field. The Swiss National Science Foundation, Grant Nr. 3.788.76, and the Swiss League of Nature Protection (SBN) have supported this study financially.

Material and Methods

Data were collected in 1978 in the Aletschwald, a nature reserve in canton Wallis, Switzerland. Ittig started ear-marking chamois in 1976. A total of 115 chamois were caught using enclosure traps and dart guns. Thirteen of these were repeats, and 5 died in the traps or due to the effect of the anaesthetic (a mixture of Rompun and Vetalar).

A herd comprising females and young was observed for 51 days from July until October. In 1978 this herd contained 60-80 chamois of which 27 were marked, thus roughly one-third of the whole herd. Positions and behaviour of focus animals, and the group surrounding them, were recorded every 15 minutes (instantaneous sampling method, Altmann, 1974). The focal animals chosen were 7 females with kids. Following the definition of Ittig, we took a 'group' to be an aggregation of chamois which were within 50 m of each other during sampling.

Study Area

The Aletschwald, where this study was undertaken, lies on the north-west side of the southern moraine of the 'Grosse Aletschgletscher'. The study area ranged from 1,800 m, at the side of the glacier, to 2,300 m above sea level on the ridge. It comprised a rectangle with 2 km sides along the glacier and ridge and 1 km across. Today the Aletschwald is a well-known tourist attraction, but visitors are restricted to marked paths. In summer

1978 it was visited by 100,000 people. The chamois are thus habituated to the presence of people and a single person may approach quite close, especially in autumn and in the evening. Nevertheless, it was difficult to follow them over long periods.

Results

Figures 10.1 and 10.2 show diurnal activity patterns for the three main behaviour categories: feeding, lying and standing/moving, for the months July/August and September/October. The data refer to all observed females and young in the herd.

The two peak feeding periods of the chamois were less marked in July/August than in September/October. They spent less time standing and moving in autumn than in summer. These two activities were concentrated in the early morning and late evening. Comparing the averages over the whole day (totals), there was a tendency in autumn towards more feeding at the expense of the other two activities. Figure 10.3 shows deviations in feeding and lying behaviour for the 7 focal females compared to the overall means. For feeding, deviations were between − 12.4 per cent and + 11.9 per cent, and for lying between − 8.6 per cent and + 4.3 per cent. The 5 per cent level was only reached for animals 310 (less feeding), 325 and 329 (more feeding), and 313 and 315 (less lying). As feeding and lying together account for more than 80 per cent of total activities, it is not surprising that those animals which were observed to feed more also spent less time lying and *vice versa*. No. 315, which fed much less but showed no increase in lying, was thus observed to spend more time standing and moving.

Spatial distributions of the 7 focal females in the months July – October are presented in Figures 10.4 and 10.5. The inhabited areas were all approximately of the same size in each month, that is: in July 27.6 ha, in August 20.8 ha, in September 26.8 ha and in October 22.0 ha. The total summer/autumn range covered 71.0 ha for the 7 focal animals and comprised a rectangle of 800 m by 1,400 m. From general observation it seemed that the home-range of the focal animals covered broadly the same area as that of the whole female/young herd. Therefore the local population density during these 4 months was 80-100 chamois per km^2. In July the animals stayed near the glacier, and as the year

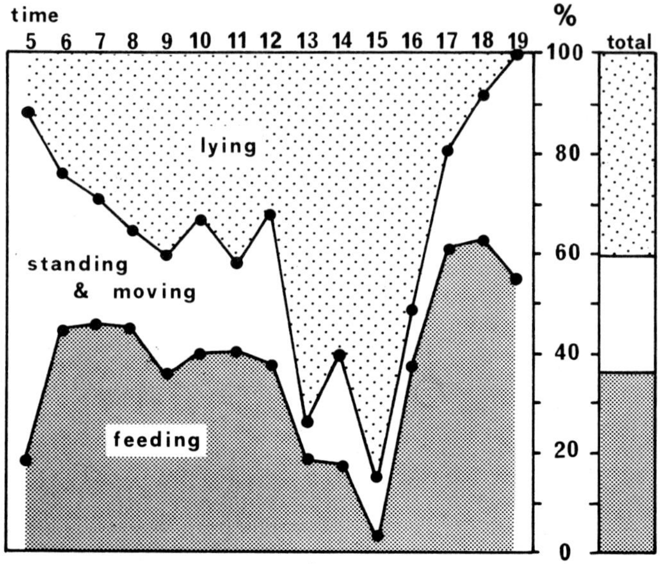

Figure 10.1: Diurnal Activity Pattern in a Herd of Female and Young Chamois in July – August. The column to the right (total) reflects the average over the whole day

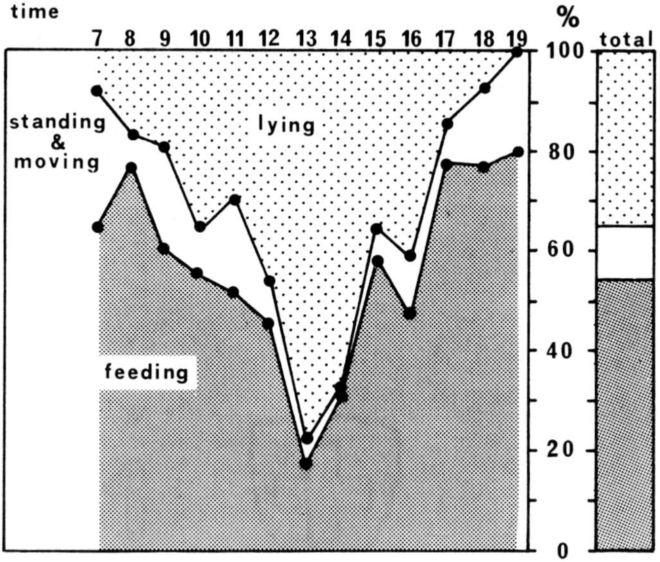

Figure 10.2: Diurnal Activity Pattern in a Herd of Female and Young Chamois in September – October. The column to the right (total) reflects the average over the whole day

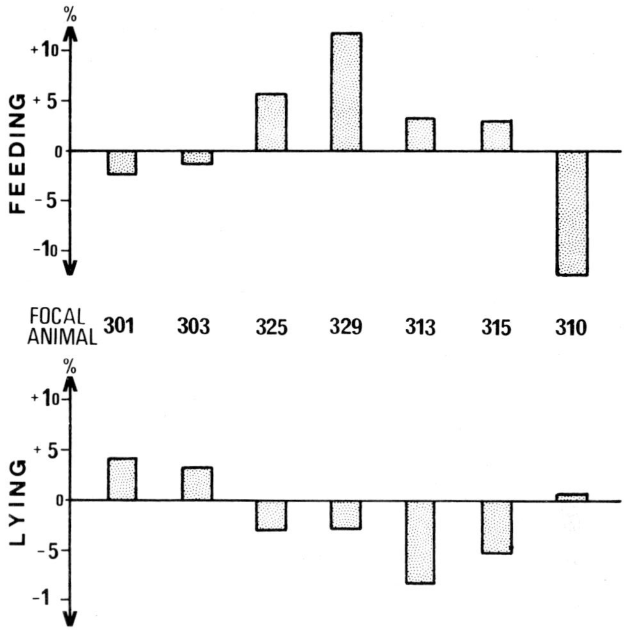

Figure 10.3: Deviations in Feeding and Lying Behaviour for Seven Females with Kids as Compared to the Overall Mean of the Herd

progressed moved east towards the edge of the moraine. During the midday rest period, from August to October, the chamois used two areas almost continuously; each of about 2 ha. These rest-areas were away from the main paths and well concealed. Individual animals chose a new lying site daily within the rest-areas.

With cluster analysis the frequencies of animals being observed in the same groups were analysed (Figure 10.6). A high correlation coefficient indicates animals that were often seen together. As expected the strongest bonds were between mothers and their kids (329-212, 313-314, 325-205, 315-331, 303-206), but there was a considerable range in the size of these coefficients. One yearling-yearling bond was the next closest relationship (319-320). Coefficients for the other relationships were all low and of a similar size. In September there was a tendency of the herd to split into two sub-herds; most animals then remained either with the western or the eastern sub-herd.

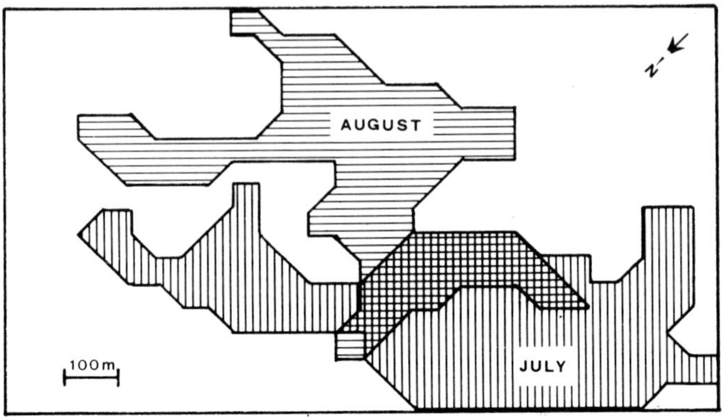

Figure 10.4: Combined Home-ranges of Seven Females with Kids in July and August

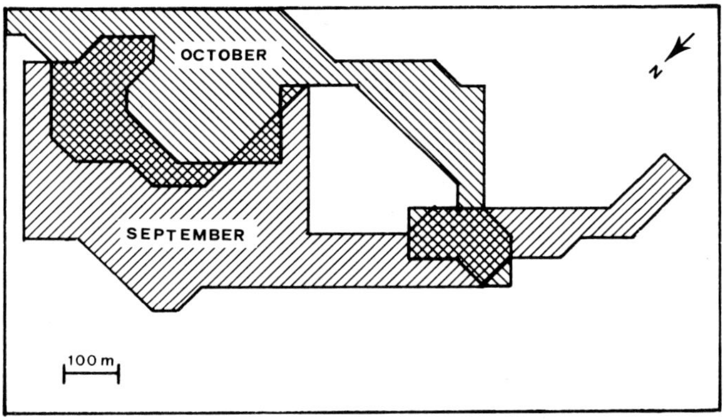

Figure 10.5: Combined Home-ranges of Seven Females with Kids in September and October

Discussion

Activity patterns of chamois in Aletschwald were less well defined than those of chamois in other areas less disturbed by man; human interference can interrupt behavioural elements. When disturbances occur during the two feeding periods, the animals must spend more of the day feeding in order to fulfil their nutritional requirements.

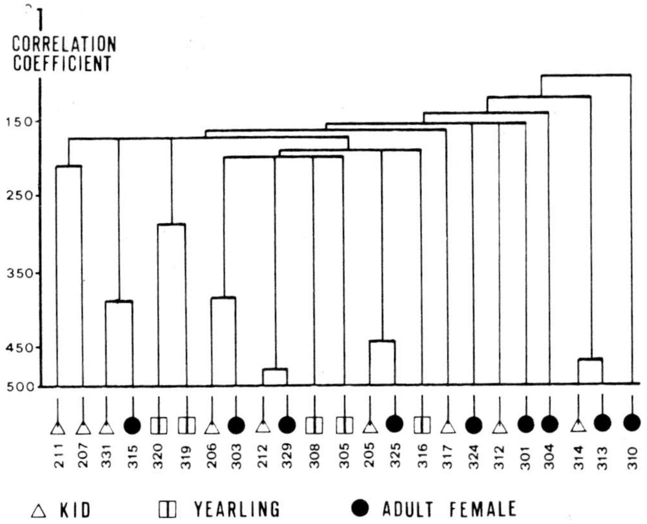

Figure 10.6: Phenogram of Marked Chamois. The correlation coefficient ranges from 0 (animals never seen in the same group) to 500 (animals always seen together)

The chamois were observed to feed more and have better defined activity patterns in autumn than in summer. As chamois are normally only active during the day, the shorter daylength in autumn allows less time for feeding; they make better use of the feeding phases than in summer and reduce the midday lying period. For the 7 focal females then there were also fewer options in addition to feeding and lying. The result that individual deviations from the general pattern were contained within a narrow range may be explained by phenomena such as group pressure, social enhancement and the necessity to fulfil nutritional requirements. To ensure valid interpretations of individual deviations, additional detailed data concerning group structure and activity are required.

Home-range size in Aletschwald was comparable to that observed by Krämer (1969a) in Augstmatthorn (Switzerland) chamois. The change in home-range between July and October was small and probably due to variations in distribution of vegetation types.

If the home-range areas for the four months are added together, the total is 26 ha greater than the actual total area used by the herd. This indicates only a small overlap between monthly home-ranges. This overlap included the two aforementioned rest-areas. The chamois progressed from the edge of the home-range to the rest-area during the morning, and back to the edge after the rest period. This explains the pattern of the behavioural category 'standing and moving' (Figures 10.1 and 10.2), and it ensures good use of the whole home-range. During the resting periods over midday, almost the same range was frequented from July till October. In contrast, the main feeding grounds, which on the whole surround the rest-area, changed from month to month (see Figures 10.4 and 10.5). It is surprising that these chamois, which can cover great distances in a short time, confined themselves to such a small area.

Results from the cluster analysis (Figure 10.6) of mother-kid bonds confirmed subjective impressions from observations in the field. Individual character traits may have accounted for differences between females. Also family relationships or possibly aspects of rank order could have influenced individual deviations. It is possible that the close bond between the two yearlings 319 and 320 originated because they were twins, although no twin kids were observed in 1978.

The weak bonds between other animals is in accordance with the literature (Krämer, 1969b, Schröder, 1971). It is probable that with more intense observations and analytical techniques more complex bonding systems could be discovered in female-young herds. The actual presence of such systems is indicated, for instance, by the split into two sub-herds in September. Refined behavioural and social studies, using many individually marked animals, are needed to further our understanding of the social structure of chamois.

References

Altmann, J. (1974) 'Observational Study of Behaviour: Sampling Methods', *Behaviour*, 49, 227-67

Krämer, A. (1969a) 'Lebensbezirke und Ortsveränderungen markierter Gemsen im Augstmatthorngebiet, Schweiz', *Z. f. Säugetierk.*, *34*, 311-15

—— (1969b) 'Soziale Organisation und Sozialverhalten einer Gemspopulation in den Alpen', *Z. f. Tierpsych.*, *26*, 889-964

Pachlatko, T. (1980) 'Aktivitätsmuster, Ortswahl, Standortswahl und Gruppenstruktur bei individuell markierten Gemsen (*Rupicapra rupicapra* L.) in zwei Geiss-Jungtier-Rudeln', Diplomarbeit, Zool. Institut, Universität Zürich

Schröder, W. (1971) 'Zur Ökologie des Gamswildes', *Z. f. Jagdwiss.*, *17*, 113-67, 197-234

CHAPTER 11

Feeding Rate and Social Stress of Female Chamois Foraging in Groups

S. LOVARI and GIOVANNA ROSTO
Istituto di Zoologia, Strada dell'Universitá 12, 43100 Parma (Italy)

Females of all ages, male kids and subadults of the Apennine chamois (*Rupicapra rupicapra ornata*) flock together to forage on the Alpine meadows from mid-summer to November (Lovari & Cosentino, in preparation). Female Apennine chamois are aggressive to each other and a clear rank order, directly correlated to age, is apparent within such feeding flocks (Locati & Lovari, in preparation). As older (and therefore dominant) females may attack and gore younger (and therefore subordinate) ones which are not quick enough to move away at their approach, a considerable tension would be expected in the latter individuals.

In this paper we report on some preliminary observations we carried out to assess the impact of social stress on the foraging efficiency of female chamois belonging to different age classes. Means were statistically compared through the two-sample *t* test (e.g. Brownlee, 1965: 297). We wish to thank A. Cederna for his help in data collection; the Abruzzo National Park Agency for providing lodging facilities and permission to work in the park; C. Scala for advice on the statistical treatment of data.

Our study area was located in the Upper Val di Rose, 1750-1942 m a.s.l., in the Abruzzo National Park (Central Apennines, Italy). Most observations were carried out during the main feeding bouts, in the late morning and afternoon, from October 5 to 12, 1983. Mean flock size was 51 individuals (SD: ± 21.4; N = 94). A variable number of ear-tagged females was present in the flock at each recording session, thus helping to assess the age of the sampled individuals. The number of bites of grass per minute was used as a measure of the rate of food intake (cf. Murton, Isaacson & Westwood, 1971) and the number of head-lifts per minute was chosen as a measure of the alertness of female chamois. As attacks by predators are rare on the open meadows, we assumed that the presence of potentially aggressive conspecifics was the main, proximate factor responsible for the chamois alert behaviour. Even if anti-predatory mechanisms were involved, these should affect both age classes we selected, thus not disturbing the significance of data. In fact, no evidence was found of different spatial positions in the flock of either age class we studied, which rules out the possibility that subordinates are particularly exposed to predation pressure (Walther, 1969, for *Gazella thomsoni*). Data were recorded through the 'focal animal' technique (Altmann, 1974) on samples of 2- to 3-year-old and 7- to 8-year-old female chamois. Each animal was kept under continuous observation for 30 minutes and its behaviour recorded in a check-sheet model every minute, for a total of 1080 observation minutes per each group.

The number of bites of grass per minute proved very significantly higher in the older (therefore dominant) females (Figure 11.1) whereas these showed a significantly lower number of head-lifts per minute (Figure 11.2) than the younger, subordinate females. Although the mean values of both parameters were not dramatically different in the two age classes (Figures 11.1 and 11.2), the consistency of each pattern may well in the long run determine a reduced food intake of the young animals, assuming an even duration of grazing bouts in both age classes. Whether this may be a key-factor in affecting winter survival, sexual maturation and reproductive success of the lower ranking age classes is still to be proven. Yet, it is suggestive that sexual maturation is greatly anticipated (1-year-old individuals may conceive) and body weight is greater in females belonging to low-density European populations of chamois (cf. Bauer, 1985) as compared with dense populations with a relatively higher degree of social stress, where first breeding occurs in 3- to 4-year-old females as in the population inhabiting our study area (Lovari, in preparation). Accordingly, it has also been shown that body weight and fertility are directly related (e.g. Albon, Mitchell & Staines, 1983).

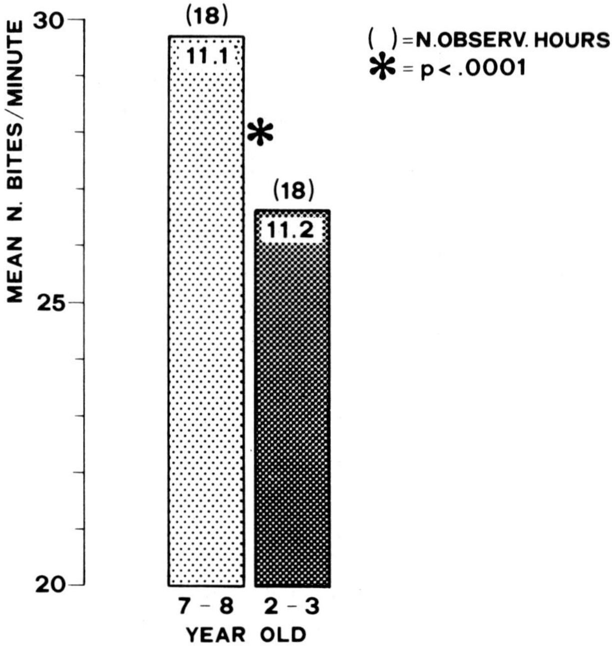

Figure 11.1: Number of Bites of Grass per Minute of Older and Younger Female Chamois. Standard deviation values are written in the upper part of the histograms

References

Albon, S.D., Mitchell, B. & Staines, B.W. (1983) 'Fertility and Body Weight in Female Red Deer: a Density-dependent Relationship', *J. Anim. Ecol.*, *52*, 969-80

Altmann, Jeanne (1974) 'Observational Study of Behaviour: Sampling Methods', *Behaviour*, *49*, 227-67

Bauer, J.J. (1985) 'Fecundity Patterns of Stable and Colonising Chamois Populations of New Zealand and Europe' in Lovari, S. (ed.) *The Biology and Management of Mountain Ungulates*, Croom Helm, London, pp. 154-65

Brownlee, K.A. (1965) *Statistical Theory and Methodology in Science and Engineering*, Wiley & Sons Inc., New York, London and Sydney

Murton, R.K., Isaacson, A.J. & Westwood, N.J. (1971) 'The Significance of Gregarious Feeding Behaviour and Adrenal Stress in a Population of Wood-pigeons *Columba palumbus*', *J. Zool., Lond.*, *165*, 53-84

Walther, F.E. (1969) 'Flight Behaviour and Avoidance of Predators in Thomson's Gazelle (*Gazella thomsoni* Günther 1884)', *Behaviour*, *34*, 184-221

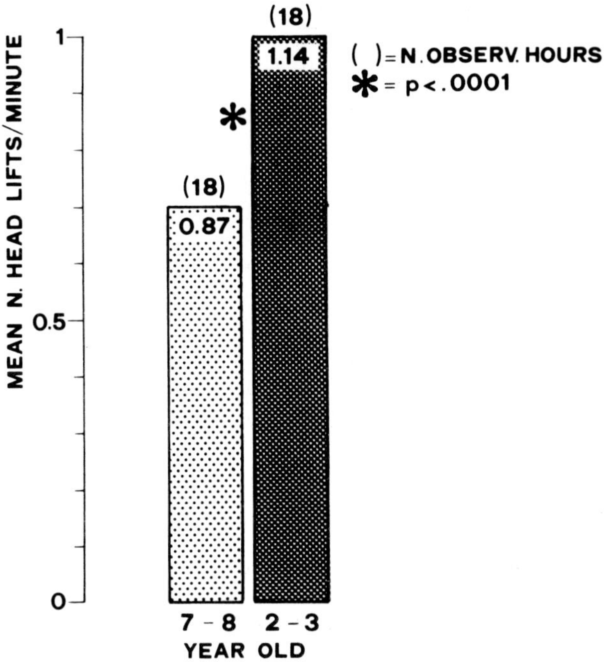

Figure 11.2: Number of Head-lifts per Minute of Older and Younger Female Chamois. Standard deviation values are written in the upper part of the histograms

CHAPTER 12

Seasonal Home Range Size and Utilisation by Female Chamois (*Rupicapra rupicapra* L.) in Northern Tyrol

J. HAMR
Department of Zoology, University of Innsbruck, Innsbruck (Austria)

Abstract

Seasonal movements and home range sizes of 11 tagged and 16 recognisable female chamois were investigated during a period of 2 years (1981-1983) inside a 50 km^2 study area. Five tagged females, aged 8 months to 11 years, were fitted with radio-transmitters. The mean size of summer-autumn ranges was 74 ha, the mean size of winter ranges was 60 ha. Different herds occupied ranges of varying sizes. Winter and summer-autumn ranges overlapped partially for some herds. The longest distance travelled between summer-autumn and winter ranges was 2.7 km. The studied animals ranged over altitudes from 950 to 1,800 m. Internal use of home ranges was highly variable during the snow-free period depending on the availability and quality of vegetation, weather and disturbance factors.

Introduction

The chamois in the study area (approximate population 450 ± 50 animals) outnumbered roe deer (*Capreolus capreolus*), red deer (*Cervus elaphus*) and a small introduced group of ibex (*Capra ibex*).

Due to the 'gregarious nature' of females (Krämer, 1969a; Schröder, 1971) large chamois herds (up to 120 individuals) require considerable areas for grazing during the snow-free periods (mid May — late November). The study area (altitude

920-2,100 m) is situated in the Karwendel Mountain Range of Northern Tyrol. It is utilised by humans both below and above the tree line, which is located between altitudes 1,500 to 1,600 m. Hikers, cross-country and alpine skiers, grazing cattle and logging operations use parts of the chamois habitat throughout the year. The densely populated valleys, busy roads and artificial dams greatly impede any long-range movements of free-living ungulates. This paper addresses the problem of chamois female habitat utilisation and seasonal migration under the described complex set of conditions.

Home range sizes of male chamois in the study area were investigated as a part of a more extensive study (Hamr, 1984). Males of different age categories used home ranges of varying sizes. Immature, one- and two-year-old males used large ranges of sizes similar to those of female herds. Young (3-5 years old) males used small (20-30 ha) summer-autumn and winter ranges, moving up and down the slope between the two. Some mature males (6-10 years old) covered considerable distances (up to 20 km) during the snow-free period. Most of these males did not return to their previous year's ranges. The seasonal ranges of older males were generally small (20-40 ha), situated below the tree-line, and varied little in consecutive years. During most of the year, male home ranges were separate from or only slightly overlapped female ranges.

Krämer (1969b) observed tagged and recognisable chamois of both sexes for a period of 2 years in the Swiss Alps. He estimated the mean yearly ranges at about 300 ha. Although Krämer mentioned a generally smaller radius of activity for males as compared to females, significant differences in yearly home range sizes between the sexes or age groups were not found in his study. The Swiss chamois were found to move mostly up and down the slopes rather than parallel to the mountain ridges.

Female chamois marked with self-attaching snare collars were observed for periods of 2 to 5 years in the Southern Alps, New Zealand. Here yearly home ranges averaged 341 ha with a maximum size of 656 ha. The mean size of summer ranges was 207 ha, mean winter range size was 70 ha. Winter and summer ranges of females overlapped. Long migrations of up to 50 km were observed mainly in male chamois (Clarke & Henderson, in press).

Funding was provided by the Foundation for Environmental

Studies (FUST) — Tyrol, The Austrian Ministry of Agriculture and Forestry and to a lesser extent by the Tyrolean Hunters' Association.

Material and Methods

Eleven female chamois were darted using a Telinject air-pressure powered gun. The darts contained a mixture of the drugs Rompun and Vetalar. The immobilised animals were tagged with coloured and numbered plastic ear-tags of the type Allflex. Five females (ages 8 months to 11 years) were fitted with Austec Electronics solar-powered, ear-tag radio transmitters. Signals were received at frequencies of 150.00-150.20 MHz using a portable, 2-element antenna and an AVM-LA 12 receiver.

Sixteen additional females, readily recognisable by broken horns or unusual pelage-markings, enlarged the sample size.

Radio as well as visual locations of experimental animals were made throughout the year, 1-6 times a week. Binoculars (7 × 42) and a spotting scope (30 × 75) were used to identify animals. All locations were recorded on forms and later transferred onto a 1:10,000 map of the study area. The map was divided into 200 × 200 m quadrats. All movements of experimental animals between locations were chronologically charted in straight lines by a computer. Various known methods of calculating home range sizes from maps or aerial photographs pose considerable problems in a highly irregular mountainous terrain. In the present study, each quadrat (area = 4 ha) containing one or more locations and/or intersected by one or more lines connecting two locations was considered a part of the animal's range (Figure 12.1).

Summer-autumn ranges were measured over the 15 May to 15 December period. Winter ranges were measured over the 1 January to 30 April period. In determining these time periods, weather and snow conditions, availability of vegetation as well as the seasonal behaviour of chamois in the study area were considered. During the remaining 4 weeks, chamois generally moved between their seasonal ranges.

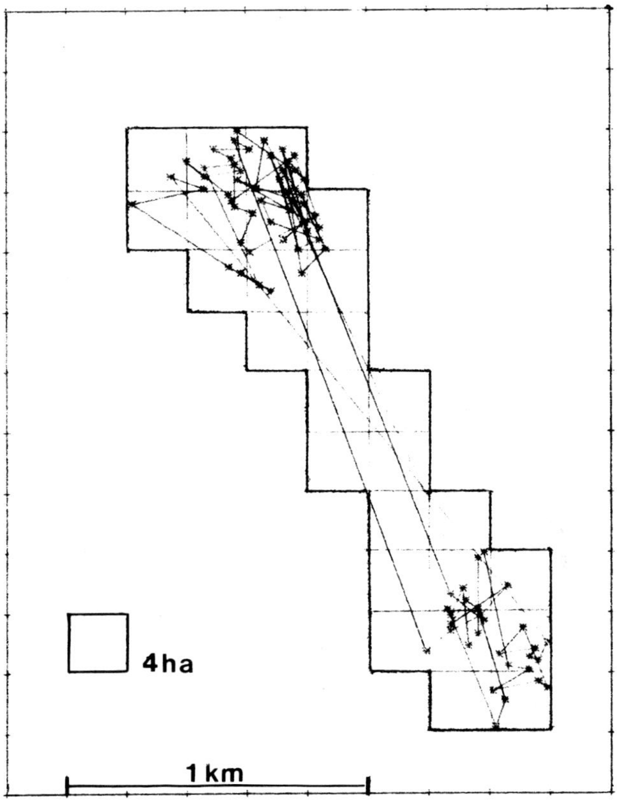

Figure 12.1: An Example Showing the Method of Home Range Size Determination. Locations of an animal are symbolised by points, movements by connecting lines

Results

Of the 27 known chamois females in the study area, 12 most frequently located females belonging to 4 different herds were used in determining home range sizes (Table 12.1). In almost 80 per cent of the transmitter locations, the experimental animal was located visually after fixing the signal. Two of the five transmitters used were lost by ripping through the ear after periods of 4 and 7.5 months respectively. Continued visual locations of these animals were possible. A transmitter-carrying female kid (age 9 months) died one month after tagging.

Table 12.1: Home Ranges of 12 Most Frequently Located Chamois Females. SFHR = Summer-autumn home range, WHR = Winter home range. Age given = age when tagged or first observed.

Animal no.	Age (years)	No. of locations (radio or visual)	Winter locations	Date of capture or first observation	Membership in herd (herd size)	Transmitter operation	SFHR (ha)	WHR (ha)	Range overlap
1	3/4	22	22	09.02.82	?	33 days	—	50	?
2	1	73	34	02.03.82	e (35-45)	14 mths	92	68	no
3	3	32	4	09.07.82	b (55-65)	——	100	—	yes
4	3	127	48	12.01.82	c (25-35)	7.5 mths	108	56	no
5	4-5	18	—	04.07.82	a (30-40)	——	32	—	?
6	5-6	24	1	11.06.82	a	——	40	—	?
7	6	45	1	22.06.82	b	——	96	—	yes
8	7	79	24	04.03.82	c	4 mths	104	60	no
9	8	26	3	05.10.81	a	——	48	—	yes
10	10	124	22	06.04.82	c	13 mths	116	64	no
11	10-11	20	1	05.06.82	a	——	40	—	no
12	12-13	27	7	02.12.81	a	——	32	—	no
Total	=	617	167						
Mean values	=	51.4	15.2				73.5	59.6	

Summer-autumn Ranges

These home ranges at altitudes of 1,300 to 1,800 m measured 32 to 116 ha with a mean of 73.5 ha (Table 12.1). The relatively small individual summer-autumn ranges of the four females belonging to herd *a* could be attributed to too few observations per animal. When all locations of these females were plotted together, the resulting summer-autumn range measured about 60 ha. It was apparent that different herds used summer-autumn ranges of varying sizes.

The summer-autumn ranges comprised larch, spruce and spruce-beech stands, alpine grasslands, talus/fellfields and cliffs (Figure 12.2). Irregular patches of knee-pine (*Pinus mugo*) and mountain alder (*Alnus viridis*) also occurred at the highest altitudes. Extreme weather fluctuations, patchiness of vegetation and disturbances caused by humans influenced the movements of chamois female herds within the ranges. The herds moved to sheltered, forested regions of their ranges during prolonged (2-5 days) rain periods or periods of strong winds (speed up to 100 km/h). In some cases, herds retreated into the forest after being hunted upon. Generally, female herds used the higher altitudes of their summer-autumn ranges for several weeks after parturition (June-July), the lower levels in August and September, and moved higher again before the rutting season (end of October).

Clearly delineated boundaries with very little overlap existed between the summer-autumn ranges of neighbouring herds (Figure 12.3). The herds ranged in number from 25 to 65 animals (females of all ages, kids and yearlings of both sexes). The number of animals in a herd remained relatively stable throughout the summer. No exchange of known females between herds was observed during the study. Immature females (2 and 3 years old) were occasionally observed outside the home ranges of the large herds. These females were sometimes seen in the company of young males or in smaller (2-5 individuals) peer groups. Some mature (5-10 years old) and 'senior' (11 years and older) females separated from their herds before the rutting season (mid November — mid December) to search alone or in small groups for male territories. Despite such movements, most females remained in the herds during the rutting time. Several males approached the herd and competed for mating rights.

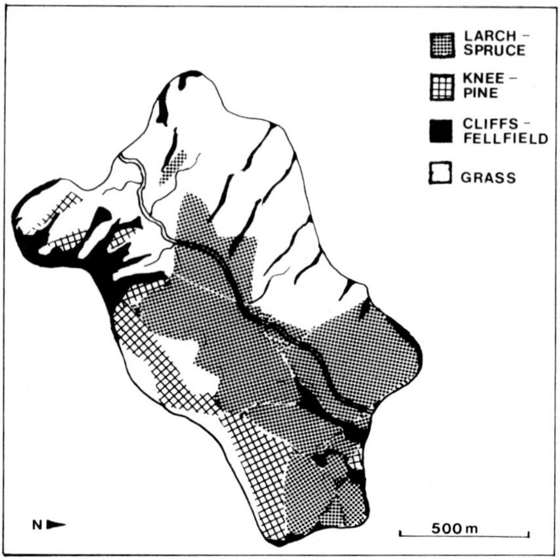

Figure 12.2: Summer-autumn Range of Herd c. Three transmitter-carrying females were in this group.

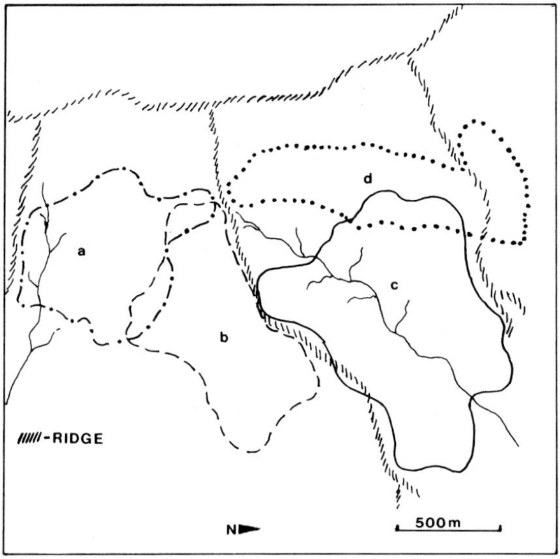

Figure 12.3: Boundary Overlap of Summer Ranges for Four Neighbouring Female Herds (the Area of Range d is a Rough Estimate)

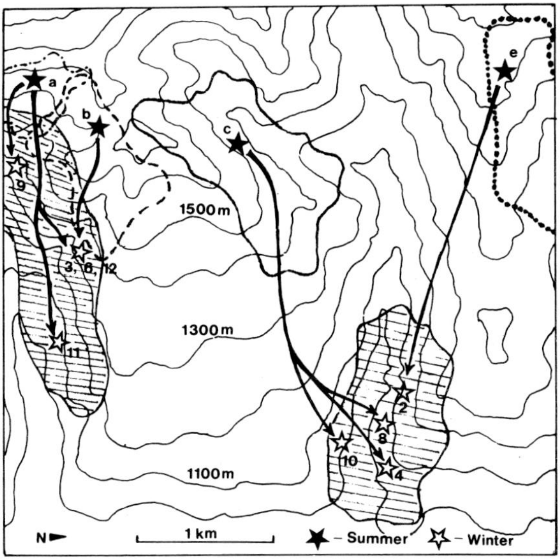

Figure 12.4: Distances Travelled between Summer and Winter Ranges for 9 Known Females of 4 Different Herds (Table 12.1). Shaded regions represent favourite wintering areas

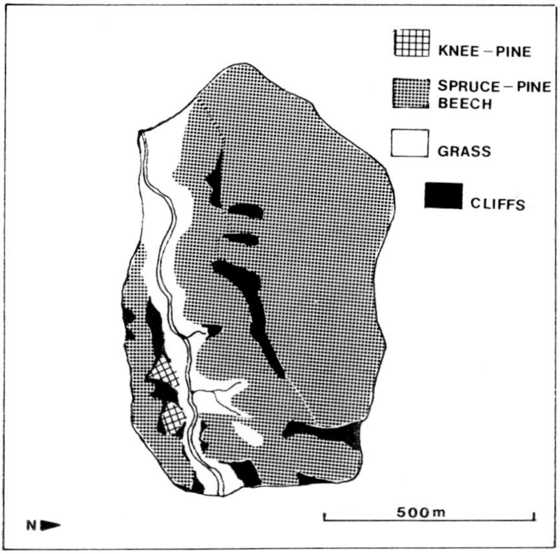

Figure 12.5: Winter Range Shared by Females of Herds c and e

Winter Ranges

Movements to lower altitudes occurred from mid-December to early January depending strongly on the depth of snow (Figure 12.4). Winter ranges measured 50 to 68 ha with a mean of 59.6 ha (Table 12.1). The studied females occupied mainly south-facing cliffs and steep grassy slopes at altitudes of 950 to 1,500 m. Open or semi-open pine-spruce-beech stands and patches of knee-pine on the cliffs were common in winter ranges (Figure 12.5). Mild winter-periods characterised by little snow on south-facing slopes enabled the chamois to exploit higher altitudes. Nonetheless, the chamois were uniformly distributed over their winter range in small groups of 2-7 individuals. Females belonging to different summer-herds shared winter ranges (Figure 12.4). Some male chamois also shared winter ranges with females of all ages.

All female chamois observed for two consecutive years were consistent in the use of their seasonal ranges.

Discussion

The results of the present investigation show the chamois could be site restrictive due to the strong influence of humans on the animals' habitat. Krämer (1969b) found larger chamois home ranges in an area with population density and environmental conditions very similar to those of the present study. His study area was less disturbed by humans.

A greater deviation of results was shown by Clarke & Henderson's (in press) New Zealand study. Much larger home ranges are to be expected in an area where habitat is relatively poor and population densities low.

Chamois strategies for habitat utilisation are very similar to those shown by mountain goats (*Oreamnos americanus*) (Smith, 1977). Most chamois females congregate on extensive summer-autumn ranges at altitudes both above and below the tree line. These ranges contain an abundance of various types of vegetation and are generally of a less rugged terrain than the winter ranges.

The quality of vegetation in various altitudinal zones of an alpine environment changes throughout the snow-free period (Atzler, 1978). In areas where cattle grazing above the tree line is extensive, the nutritional quality of ground vegetation shows two peaks: (a) in July and the beginning of August; (b) from the end of September through most of October. The observed utilisation of summer-autumn ranges by herds of females shows adaptations to altitudinal differences in vegetation quality and distribution. The size of female summer-autumn ranges thus seems to depend on the quality of the range as well as on the extent of ranges of neighbouring herds rather than on the actual herd size.

The robust young males, mostly in smaller groups above the tree line, have in comparison with the lactating females limited nutritional requirements and consequently use smaller home ranges. They also remain above the tree line during prolonged windy or rainy periods. Solitary old males — mostly below the tree line, require only a small area to fulfil their nutritional requirements.

During winters with deep, persistent snow, which are typical for the study area, herds disperse to form small foraging groups on the smaller winter ranges thus maximising the use of available food and reducing intraspecific competition. Slopes and cliffs chosen for winter ranges shed snow more readily than other terrain types. Forage is therefore more accessible and snow shallower. Suitable winter habitats are relatively few and therefore attract females and males from different summer ranges.

The observed consistency in the use of home ranges was supported by observations of known females by game wardens over periods of up to 7 years. Traditional use of home ranges by females has also been reported by Rideout (1977) for mountain goats and by Geist (1971) for bighorn sheep (*Ovis canadensis*). No emigration of females from the area was observed during the study. The home range use exhibited by reproductively mature females in the Karwendel Mountain Range suggests home range constancy. Significant differences in home range size or use between females of different age categories were not apparent. Frequent emigration of mature males apparently ensures the exchange of genes between chamois herds or sub-populations.

References

Atzler, R. (1978) 'Schwankungen der Qualität der Bodenvegetation während der

Vegetationszeit im Rothirsch-Untersuchungsgebiet Ammergauer Berge', Diplomarbeit am Fachbereich Forstwissenschaft, Abteilung für Wildforschung und Jagdkunde der Universität München

Clarke, C.M.H. & Henderson, R.J. (in press) Home Range Size and Utilization by Female Chamois (*Rupicapra rupicapra* L.) in the Southern Alps, New Zealand. *Acta Zool. Fenn.*

Geist, V. (1971) *Mountain Sheep — A Study in Behaviour and Evolution*. The University of Chicago Press, Chicago & London

Hamr, J. (1984) Home Range Sizes of Male Chamois (*Rupicapra rupicapra*) in the Tyrolean Alps, Austria. *Acta Zool. Fenn. 171*, 293-8

Krämer, A. (1969a) 'Soziale Organisation und Sozialverhalten einer Gemspopulation (*Rupicapra rupicapra* L.) der Alpen', *Z. f. Tierpsychol.*, *26*, 889-964

—— (1969b) 'Lebensbezirke und Ortsveränderungen markierter Gemsen im Augsmatthorngebiet, Schweiz', *Z. f. Säugetierk.*, *34*, 311-15

Rideout, C.B. (1977) 'Mountain Goat Home Ranges in the Sapphire Mountains of Montana', *Proc. First International Mountain Goat Symposium*, Kalispell, Montana, pp. 201-11

Schröder, W. (1971) 'Zur Ökologie des Gamswildes (*Rupicapra rupicapra* L.)', Dissertation, Universität Göttingen, FRG

Smith, B.L. (1977) 'Influence of Snow Conditions on Winter Distribution, Habitat Use and Group Size of Mountain Goats', *Proc. First International Mountain Goat Symposium*, Kalispell, Montana, pp. 174-89

CHAPTER 13

Seasonal Fluctuations in the Spatial Distribution of Chamois and Moufflons on the Carlit Massif, Pyrenees

G. GONZALEZ

Laboratoire de Neuroethologie, Université Paul Sabatier, 31000 Toulouse (France)

Abstract

The area occupied by the Pyrenean chamois (*Rupicapra rupicapra*) and the moufflon (*Ovis musimon*) on the Carlit mountain range (25,000 ha) was plotted from notes made of sightings and footprints during a two-year period (1981-1982). The Pyrenean chamois are fairly evenly distributed, as small groups of animals inhabit the higher zone throughout the year. The moufflons, on the other hand, from November until April group together on refuge slopes facing south but are widely scattered during May and June. Some areas are inhabited by both species throughout the year. Observation of individual animals by means of natural or artificial markings made it possible to map out their respective home ranges. On the whole, the chamois' home range was smaller than that of the moufflons. Moufflons migrate in spring. The spatial overlapping of these species did not exceed 10-18 per cent of their total home range. There was very little overlapping during winter, with a clear altitudinal segregation due to reciprocal intolerance.

Introduction

Boillot (1980), Corti & Leonard (1981) and Delaunay (1981) have already described how chamois extend their home ranges in spring, colonising areas where the snow has melted. Comparable behaviour has been observed by Pfeffer & Settimo (1973) and Cugnasse (1982) in moufflons. On the other hand, Filej (1976) provided some data showing that the introduction of a population of moufflons in Ursja Gora had a negative effect on the chamois there.

In this preliminary report, I will describe the seasonal fluctuations in the spatial distribution of populations and individual animals belonging to both species in an area where they coexist.

I wish to thank the hunting associations which gave me their assistance, and also Mr Claude Novoa for his friendly help in the field. Finally, I am also grateful to the Office National de la Chasse for their financial support for this study.

Material and Methods

The geographical distribution of the two species was mapped (1:50,000 and 1:25,000) by noting down the position of the animals observed (binoculars 10 × 40; telescope 25-60 × 80) during reconnaissance hikes (25,000 ha) from January 1981 until January 1983. Some animals (5 Pyrenean chamois, 5 moufflons), morphologically recognisable (coat colour, horn ridges) or artificially marked by a collar, were kept under observation. The space occupied by a population or by a marked animal was plotted using the 'convex polygon method'. Slope inclination was not considered in the estimate of the range extension. Thus, values for the home ranges are tentative and give only an approximate estimate of the temporal variations of spatial distribution.

Study Area

The study area (Figures 13.1 and 13.2) is a granite mountain massif (1,400-2,900 m a.s.l). Thirty per cent of this massif is covered with forest, and is permanently or temporarily inhabited by both species. The rest of the study area consists of Alpine meadows interspersed with rocks and cliffs. The snow cover lasts from December to April and regions made inaccessible by winter snowfalls could not be properly explored.

Figure 13.1: The Study Area

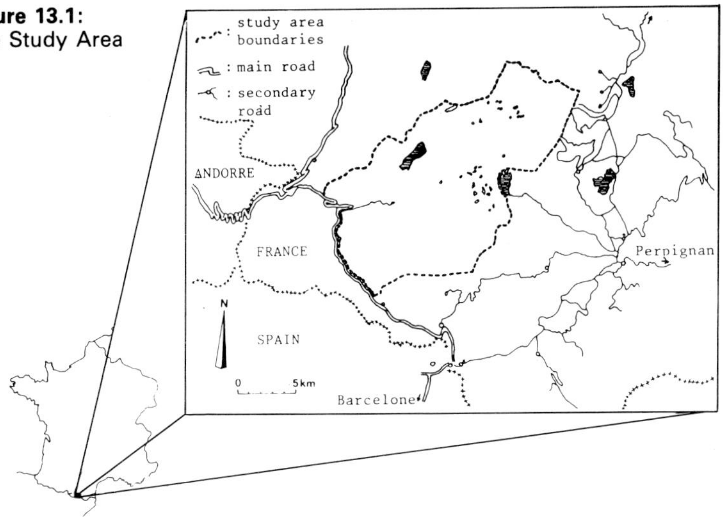

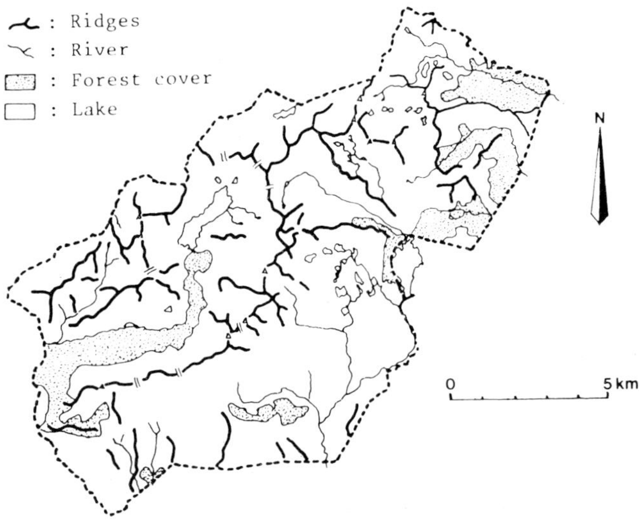

Figure 13.2: Topography of the Study Area

Results

Estimates of the Areas Frequented by Each Species over Different Periods of Time

Figure 13.3 shows the variations in the home ranges of the two species. The dispersion of animals was minimal in the winter months (November-December and January-February) and maximal in May-June and September-October. The north-eastern region (Figure 13.3) was used only by chamois. This was also true of the south-eastern border of the area which was inhabited by chamois throughout the year. The central portion of the study area (at its highest altitudes) was only inhabited when snowfall was minimal (May to October). It was very difficult to search the area during winter, but some excursions made on skis would confirm that the area was deserted at that time. From January to April almost the whole population of moufflons kept to a 800 ha south-facing slope where the snow melted more rapidly, on the western border of the study area. Some of the moufflons, mostly males, spent the winter in a valley on the south-eastern end of the massif where the snow is now very deep and melts rapidly in spring; the moufflons desert these areas completely in July and August. On the other hand, half the winter population of chamois (*i.e.* approximately 400 individuals) stay on the same south-western slopes all the year round. Home range variations based on animal sightings indicate that chamois occupied the study area more constantly; there was no scattering of the population on the winter range or summer concentration on higher crests. By contrast, the moufflons occupy an area which varies according to the seasons with a maximum extension in May and June. The area where the two species coexist varies in the same way, but it never exceeds 10-18 per cent of the total home range of both species.

Spatial Distribution of Individually Identified Animals

The areas where identifiable moufflons were sighted varied with the seasons, which confirms what I observed for the population. South-facing slopes were frequented during the snowfall months (November to June). The higher crests and peaks were inhabited by the identifiable moufflons only when snowfall was low or absent (June to October). Finally, land at a lower altitude, with

Figure 13.3: Evolution of the Approximate Home Ranges of Chamois and Moufflons from January 1981 to January 1983

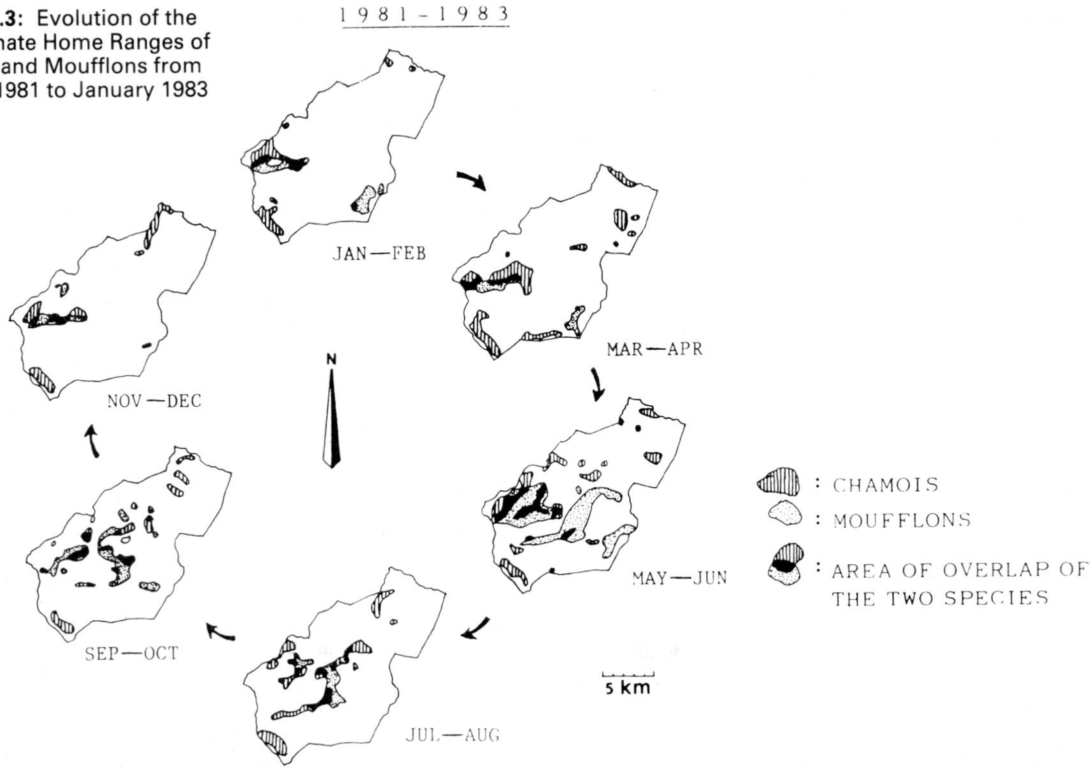

pastures, where the snow melts early in the summer, proved very attractive to the moufflons during May.

Estimates of chamois home ranges revealed a different spatial distribution. All areas were small (20-100 ha): male no. 1 was seen from November 1981 until November 1982 in an area of 30 ha; female no. 1 apparently made a short migration of about 1 km between April and July-August. Female no.2 was observed on the south-west slopes of the study area in July 1981 and November 1982; she probably belonged to the resident population. Male no. 2 was not in his usual area between June 1982 and September 1982, but returned to it at the end of September.

Discussion

The home ranges of the moufflons and chamois populations, as well as those of the identifiable animals of both species, differ during the year. Whereas all moufflons and three-quarters of the chamois spend the winter months on south/south-western slopes, where the snow melts rapidly, the overall spatial distribution of the chamois seems more even because some small groups stay at the same altitude all year long. These areas are mostly ridges where the snow is cleared by violent winds. Chamois are on the whole less gregarious than moufflons, so their distribution is relatively uniform. Furthermore, moufflons have much larger home ranges than chamois. In spring, the chamois home range is approximately twice as large as in other seasons whereas the moufflons home range increases to ten times the winter extension. Chamois seem to be fairly sedentary, sometimes staying in the same area throughout the year.

My results suggest that moufflons of all ages migrate from their winter areas to the higher summer ranges (2,200 to 2,900 m) in the month of May. Then they have to travel through mountain passes (Col Rouge and Porteille du Bac d'Ourteille) in search of new pastures. This migration seems paradoxical as at this time new grass shoots are also growing in the area they abandon. In fact, the main cause of this migration could be temperature increase rather than food requirements.

In the region where the two species coexist, there is an altitudinal segregation due to an apparently reciprocal intolerance: the moufflons stay at lower altitudes than the chamois; this is also described by Pfeffer & Settimo (1973) with regard to the Mercantour reserve. Nevertheless here we have one element which differs from previous observations; the spatial distribution of the two species partly overlaps in Carlit. The shared area is maximal in May and June when the moufflons' range is at its maximum. Then it gradually decreases, but in November-December it increases again and builds up to 15-18 per cent of the study area. At this time the moufflons inhabit a territory that is one-tenth of the size of that occupied in May-June. The area in common is minimal in July-August, as temperature is apparently one of the factors involved in the segregation of the two species and, in these months, chamois move towards the shady northern slopes or the forested areas.

References

Boillot, F. (1980) 'Différents Types de Marquages Utilisés pour L'étude Écoéthologique du Chamois Vosgien', *Bull. Off. Nat. Chasse*, *35*, 13-18

Corti, R. & Leonard, P. (1981) 'Dépouillement des Observations en Continu des Gardes sur le Chamois dans le Réserve du Mercantour de 1975 à 1978', *Bull. Off. Nat. Chasse*, *50*, 32-6

Cugnasse, J.M. (1982) 'Résultats Préliminaires sur les Observations de Mouflons Individualisés à L'aide de Colliers (Caroux)', *Bull. Off. Nat. Chasse*, *59*, 32-6

Delaunay, G. (1981) 'Etude et Gestion des Populations de Chamois dans le Parc National des Ecrins', Thèse de 3° cycle. Université de Rennes

Filej, J. (1976) 'Influence des Colonies de Mouflons sur la Population de Chamois dans l'Ursja Gora', *II° Symposium chamois*, Bled (translation from German by C. Gindre, Off. Nat. Chasse), (mimeographed), pp. 1-6

Krämer, A. (1969) 'Organisation et Comportement d'une Population de Chamois des Alpes', *Z. f. Tierpsychol.*, *26*, 889-964 (translation from German by C. Gindre, Off. Nat. Chasse)

Pfeffer, P. & Settimo, R. (1973) 'Déplacements Saisonniers et Compétition Vitale entre Moufflons, Chamois et Bouquetins dans la Réserve du Mercantour (Alpes Maritimes)', *Mammalia*, *37*, 203-19

CHAPTER 14

Male and Female Alpine Ibex: Phenology of Space Use and Herd Size

F. FRANCISCI,[1,3] S. FOCARDI[2] and L. BOITANI[1]

1. Dipartimento di Biologia Animale e dell'Uomo, Università di Roma, 00185 Roma (Italy)
2. Istituto Nazionale di Biologia della Selvaggina, via Stradelli Guelfi 23/a, 40064 Ozzano Emilia (Italy)
3. *Present Address*: Istituto Nazionale di Biologia della Selvaggina, via Stradelli Guelfi 23/a, 40064 Ozzano Emilia (Italy)

Abstract

During a two-year pilot study, the distribution of the ibex, *Capra ibex* (L.), was investigated in an area of the Gran Paradiso National Park, Italian western Alps. Results obtained by regression analysis between male and female occurrences per study area units provided evidence for periodical, spatial segregation between sexes outside the rutting season. Further analysis has shown a gradual variation from winter range separation to avoidance on the same range as a result of daily different space utilisation by sexes during summer. Moreover sex-related differences are found in the periodical shift of the mean altitudinal level and in the preference for slopes with different gradients.

Introduction

Since the Gran Paradiso National Park was established in 1922, annual ibex censuses have been carried out by Park wardens. Traditionally, such censuses are conducted before and after the kidding season. The analysis of these data from 1945 onwards (Linzi, 1979) shows that the total population has settled at about 3,500 individuals since 1962. The two-year pilot study was designed to assess the main factors regulating the only autoctone alpine ibex population surviving in Europe.

Recent studies on Caprinae ecology have shed light on a variety of aspects related to segregation between sexes in species inhabiting northern climates (Nievergelt, 1966, 1974, 1981; Geist, 1971; Geist & Petocz, 1977; Schaller, 1977). Because of limited food resources it is adaptive for both sexes to have different diets, e.g. a reduced food competition during the gestation period (winter) would benefit the fetus as females may have more energy to invest in its growth (Geist, 1974).

Marked sexual dimorphism in ibex has been correlated to different habitat use, with the males showing less clearcut preferences than the smaller, ecologically more sensitive females (Nievergelt, 1981).

The present study was financially supported by the Gran Paradiso National Park Agency. We wish to thank C. Pulcher for helping in winter fieldwork, P. Molinari for data processing and S. Lovari for his suggestions.

Study Area

The study area is roughly a 5,346 ha (plane surface) rectangle with its major axis oriented north-south. Altitudes ranging from 1,500 to 3,260 m are included. The eastern and northern limits of the study area coincide with the right slope of the Nontey valley and the Eyvvia river, below the Nontey river junction. The western limit of the study area is bordered by a series of small glaciers whereas the south boundary consists of the Grand-Croux and Money glaciers. The Nontey valley, running southwest-northwest, is the main axis of the study area. Tributaries from the left slope of the valley have carved a number of smaller valleys oriented west to east; all of these are important ibex access routes to high altitude pastures. Outstanding among others is the upper Lauson valley: a wide, high altitude 'bowl' of continuous, unbroken pastures.

Deep narrow periglacial valleys, carved in metamorphic gneiss rock, characterise the landscape of the study area. Groundcover is poor; sparse stands of larch trees (*Larix* spp.) may reach the 2,500 m contourline, but only the north-northeast section of the study area has an extensive larch wood cover. Except for *Alnus* spp., along the upper Nontey river, the greater part of the study area is covered with a variety of alpine grasses including *Carex* spp., *Festuca* spp., *Poa* spp. and *Juncus* spp. On average, 300 ibex are censused in the study area each year.

Material and Methods

The study area was divided into a grid of 220 quadrats, each measuring 25 ha.

During the period July 1979 — July 1981, 4069 male and 1739 female ibex were observed. Such a biased sex-ratio was due to non-random sampling of the study area because of the difficult access to the preferred ranges of the females. A correction was introduced by dividing the ibex number in each quadrat by the number of visits in each month. The resulting values give a sex ratio (male/female) equal to 1.1/1, which is comparable with the 1/1.2 reported by Linzi (1979). Only class I to IV males (Geist, 1971) were considered while female groups included adult females, kids and yearlings of both sexes.

The mean value of group size for males ($\bar{X}$) and females ($\bar{Y}$) was computed for each month and for the whole period of observation.

The monthly mean altitudinal level for males (h_x) and females (h_y) was computed according to the relations:

$$h_x = \frac{\sum_{i=1}^{220} \sum_{t=1}^{30} x_{i,t} \cdot h_i}{\sum_{i=1}^{220} \sum_{t=1}^{30} x_{i,t}}$$

$$h_y = \frac{\sum_{i=1}^{220} \sum_{t=1}^{30} y_{i,t} \cdot h_i}{\sum_{i=1}^{220} \sum_{t=1}^{30} y_{i,t}}$$

where $x_{i,t}$ and $y_{i,t}$ represent the observations of male and female groups in each grid unit (i = 1, 220) at each observation time (t = 1, 30), every month; h_i is the median of the altitudes of the quadrats with ibex occurrence.

The linear regression between the numbers of female and male per quadrat was computed at monthly intervals. Two different procedures to calculate the regression line were used: (1) using the $x_{i,t}$ and $y_{i,t}$ values (regression coefficient b); (2) using the mean monthly values of males and females for each quadrat (correlation coefficient b'). The coefficient b informs about the short-time association between male and female groups: a positive value indicates their association in space and time. Conversely, coefficient b' reveals the use of the same range, during a month, disregarding synchronous association between the sexes on the same area unit. Finally the y-intercept (a), derived from the first regression line, informs on the degree of overlap between male and female ranges: when positive, females occurred over a larger area than males did. Analysis of the regression of females on males has been used for descriptive purposes only: no attempt was made statistically to test the periodicity of b, b' and a. The difference between the b and b' values and 0 was statistically evaluated by the 't' test (Sokal and Rohlf, 1981). The same analysis performed by use of the correlation coefficient yields quite similar results.

Unfortunately, it was impossible to collect enough quantitative data in October. Therefore this month was not considered in the analysis.

Results

The monthly variation of b (Figure 14.1) is characterised by positive values during the rutting season (December — January) whereas during late winter-early spring (March) such values reverse abruptly. A smooth increase can be observed through spring and summer. When b' is considered, a similar positive peak during the rutting season can be detected, but a second positive peak is also apparent in summer (Figure 14.2). A comparison between Figures 14.1 and 14.2 suggests that both sexes relied on the same area during the warm season, although no short-time association occurred ($b = 0$ in August). In spring, negative b values, indicating a separation between sexes, are determined by the use of different parts of the study area (b' equal to zero).

Variations in a (Figure 14.3) indicate that in summer (June-August) the area where females occurred was larger than that of males. Therefore, since both sexes were found on the same areas (Figure 14.2), the aggregation level would seem to be different. Conversely, in September group size is quite similar in both sexes (Figure 14.5) and the high a values are maintained by separation

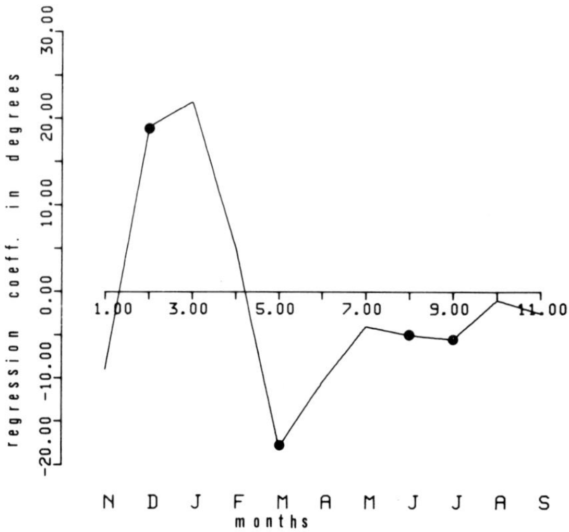

Figure 14.1: Regression Coefficient *b* Plotted as Function of Time, *b* values expressed in degrees. Dots indicate *b* values statistically different from zero ($p < 0.05$). *b* is an index of the synchronous occurrence of male and female ibex on the same quadrat.

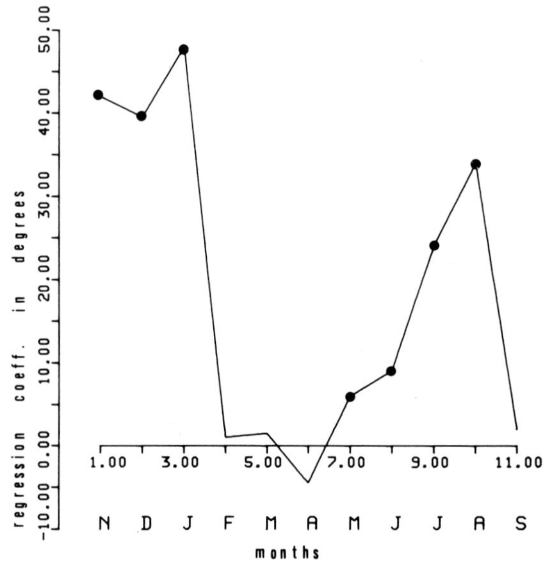

Figure 14.2: Regression Coefficient *b′* Plotted as Function of Time. *b′* values expressed in degrees. Dots indicate *b′* values statistically different from zero ($p < 0.05$). *b′* is an index of the use of the same quadrat by both sexes, averaged over a month interval

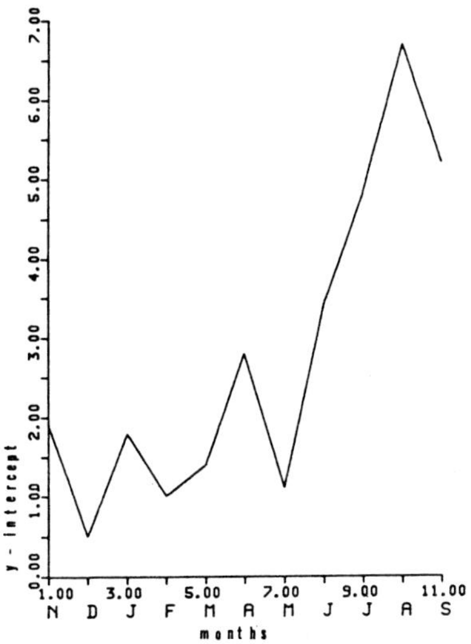

Figure 14.3: The y-intercept *a*, referred to the correlation coefficient *b*, plotted as function of time. *a* is non-dimensional (see also Material and Methods).

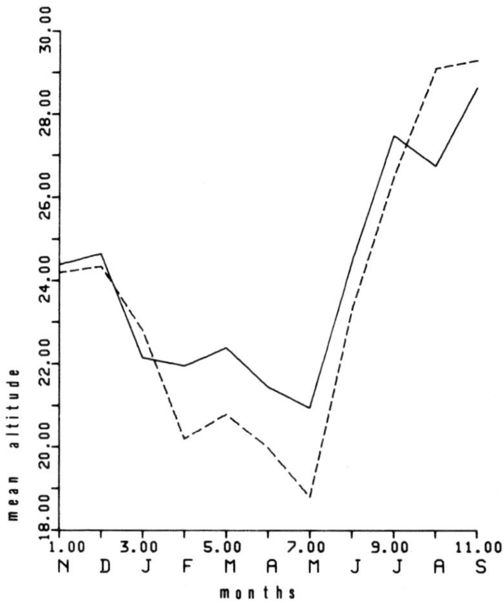

Figure 14.4: The Mean Altitude (in metres) Plotted as Function of Time. Ordinate: metres by 100. Continuous line, females; dashed line, males

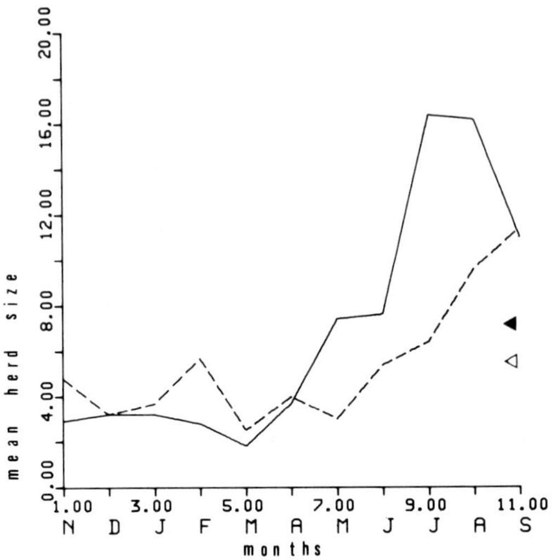

Figure 14.5: The Mean Herd Size (number of individuals) Plotted as Function of Time. Continuous line, males; dashed line, females. Arrows represent mean herd size computed for the whole study. Males, solid arrow; females, open arrow

of ranges (Figure 14.2). When b is negative but greater than $-45°$ (March and April) the y-intercept is smaller than the x-intercept showing that males occupied larger sectors of the study area than females did. Both sexes, while showing the same temporal patterns of altitudinal variations, reached the highest altitudes in summer and autumn, whereas during winter and spring they moved to lower levels (Figure 14.4). Females had narrower seasonal altitudinal shifts than males (difference of means for September and May: 772 m and 1,039 m, respectively). Such differences were caused by altitudinal separation at lower levels: males, from February through May, were always found lower than females (mean diff.: 206 m) whereas both sexes reached almost the same high altitudes in September. Negative b and null b' values for such periods may be related to the altitudinal separation between the sexes.

The reciprocal influence exerted by habitat condition and social behaviour can be appreciated considering the variations of the mean herd size (Figure 14.5). From the beginning of the rutting season through April, the herd size was similar in both sexes, although females showed slightly larger groups than males. In May the mean herd size of males increased at a higher rate than that of females. After a peak in July and August the male herd size sharply declined to the same value observed for females in September. The different social organisations of males and females in summer may be dependent on the vertical movement of the ibex, which is positively influenced by the local topographic features, more open and less broken at the higher altitudes. As the animals gain in altitude larger groups can thus be formed. Figure 14.2 shows that in summer the sexes range over the same area, so that the corresponding *a* values (Figure 14.3) may then be caused by the different grouping patterns of the sexes. The more scattered females, in fact, were distributed over larger parts of the study area than males were.

Interestingly, the variation in the use of slopes with different

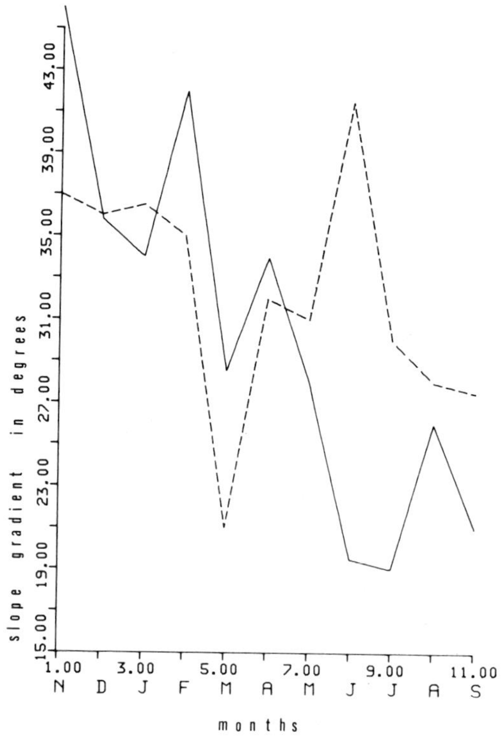

Figure 14.6: Mean Slope Gradient Plotted as a Function of Time. Continuous line, males; dashed line, females

gradients (Figure 14.6) shows considerable divergence in relation to the kidding season (June-July): unlike males, females preferred steep slopes and cliffs.

Discussion

These data suggest that different spatial habits may be used by male and female ibex to withstand the severe alpine climate. Large herbivores satisfy their food requirements by both increasing food intake and using a greater fraction of the available biomass than smaller species do (Clutton-Brock & Harvey, 1983). Consequently they are more efficient at subsisting on low quality vegetation.

Female ibex weighing about only half an adult male can survive with lower energy requirements (Couturier, 1962). However, during lactation, their food requirements are expected to increase greatly. Moreover, the energy expenditure of females, while moving across broken and steep terrain, is lower than for males, allowing them easy access to the sparse vegetation found in such areas.

While both sexes, during rutting season, show similar spatial patterns, in February several of the investigated parameters (b, b', h_x, h_y) vary sharply according to different spatial behaviours of the sexes. In winter and spring, males reduce their energy expenditure by 'drifting' to lower altitudes while females maintain higher altitudes. Thus, females may exploit better food resources than males because of the favourable thermal and snow conditions of the higher slopes.

Later on, at the onset of the melting season (May) the flush of new grasses permits all males to aggregate in large herds over the open meadows (Nievergelt, 1966).

Because of the oncoming kidding season females do not show a similar pattern. In fact they retreat to steep, well-protected areas, where food is scattered and large aggregations are impossible. The increasing food requirements during lactation may force the females to move to the open meadows.

Although segregation between the sexes is most obvious during winter, when separate ranges are maintained, mechanisms of separation may also be active during summer. In the high-altitude summer ranges reduced spatial segregation is compensated by increasing differences in aggregation patterns, requirements in habitat geometry and, possibly, daily activity rhythms.

In conclusion habitat use by males is more opportunistic than by females. Depending on body size and related energy requirements, males roam over larger areas then females, grazing on short-lasting plant communities with high standing crops. As a consequence, because of the interplay between movement pattern, distribution of food resources and topographical constraints, the attained sociality level is different in the sexes: while female social milieu is relatively stable throughout the year, males experience strong periodical variations.

References

Clutton-Brock, T.H. & Harvey, P.H. (1983) 'The Functional Significance of Variation in Body Size Among Mammals' in J.F. Eisenberg & D.G. Kleiman (eds.), *Advances in the Study of Mammalian Behavior*. Special publication no. 7. The American Society of Mammalogy, Shippensburg PA, pp. 632-58

Couturier, M. (1962) *Le Bouquetin des Alpes*. Arthaud, Grenoble

Geist, V. (1971) *Mountain Sheep*. Chicago University Press. Chicago & London

—— (1974) 'On the Relationship of Ecology and Behavior in the Evolution of Ungulates: Theoretical Considerations' in V. Geist & F. Walther (eds.), *The Behavior of Ungulates and its Relation to Management*, IUCN Publ. no. 24, Morges, pp. 235-47

Geist, V. & Petocz, R.G. (1977) 'Bighorn Sheep in Winter: do Rams Maximize Reproductive Fitness by Spatial and Habitat Segregation from Ewes?' *Can. J. Zool*, *55*, 1802-10

Linzi, M.P. (1979) 'Analysis of the Gran Paradiso NP Ibex Census Records', Unpublished Doctoral Thesis, University of Rome (in Italian)

Nievergelt, B. (1966) *Der Alpensteinbock*, Mammalia Depicta, Parey-Verlag, Berlin

—— (1974) 'A Comparison of the Rutting Behavior and Grouping in the Ethiopian and Alpine Ibex' in V. Geist & F. Walther (eds.), *The Behavior of Ungulates and its Relation to Management*. IUCN Publ. no. 24, Morges, pp. 324-41

—— (1981) *Ibex in an African Environment*, Springer-Verlag, Berlin

Schaller, G.B. (1977) *Mountain Monarchs*, Chicago University Press, Chicago & London

Sokal, R. & Rohlf, J. (1981) *Biometry*, Freeman & Co., San Francisco, pp. 859

CHAPTER 15

Group Size and Composition of the Spanish Ibex (*Capra pyrenaica* Schinz) in the Sierras of Cazorla and Segura

CONCEPCIÓN L. ALADOS
Estaciòn Experimental de Zonas Aridas, 1 General Segura Str., Almería (Spain)

Abstract

The study was carried out in the Sierras of Cazorla and Segura, in the southeastern part of the Iberian Peninsula. Field work lasted one year (from February 1982 to January 1983). For each observation the size and composition of the group and the characteristics of the habitat were recorded. The rut in the population of ibex under study begins earlier than for the alpine ibex (*C. ibex*) and Walia ibex (*C. waliae*), and finishes earlier. In January and February the mixed groups begin to break up into male and females-kids groups. From April to October herds of juveniles (of both sexes) form, during which time the females give birth and rear their newborn. Groups of juveniles and of females-kids predominate in habitats of high vegetation density, mixed groups being more frequent in habitats of sparse vegetation. During the spring and autumn, in their movement to and from the rutting grounds, the male groups occur more frequently in areas of more dense vegetation. Group size is usually larger in open areas than in those of denser vegetation, mixed groups being the largest. With regard to the males attached to those mixed groups, the younger their age the longer they remain in them. Finally, the considerable intensity of tourist pressure on the area during the summer months causes a break-up of the groups of females with kids, as the animals have to move away and hide in areas of denser vegetation, where food is of poorer quality.

Introduction

The Spanish ibex (*Capra pyrenaica*) lives in isolated populations distributed throughout the Iberian Peninsula. This study concentrated on the population of the Cazorla and Segura mountain ranges, the animals being currently ascribed to the subspecies *C.p.hispanica*.

The size and composition of social groups are important features of social structure. Referring to this, Jarman (1974) establishes five categories of social groups for African antelopes, demonstrating a relationship between feeding behaviour and group size, e.g. grazing species, to a great extent unselective, may be in large groups as the grass on which they feed can bear much pressure, whereas the browsing species, which are much more selective concerning their diet, live in small groups or are solitary.

As to the advantages of living in a group, most authors (Brock & Riffenbergh, 1960; Cushing & Harden Jones, 1968; Hamilton, 1971; Vine, 1971; Altmann, 1974; Bertram, 1978; Siegfried, 1980; Lipetz & Bekoff, 1982; Underwood, 1982) agree with the idea that individuals in a group reduce their vulnerability to predation through an increase in vigilance. Here Jarman (1974) adds the idea that herds can sometimes exceed their optimum size, thus losing the advantage of alarm communication when they are attacked.

It is worth remembering, however, that a given species can show a distinct social strategy according to the habitat occupied. Thus, solitary species such as roe or white-tailed deer, on the rare occasions in which they occupy open habitats, often group together in herds (Geist, 1974b).

In general, the members of the Caprinae subfamily are highly gregarious species, living in herds of females with kids and young or in all-male groups. This segregation of the sexes in adults is only broken during the rut, when they form mixed groups of both sexes and all ages. However, there are exceptions in the most primitive taxa, i.e. the serow and the goral (Schaller, 1977).

Groups of young animals of both sexes occur in the Caprinae during the period when females are giving birth, as also happens in the alpine ibex (Couturier, 1962), Himalayan tahr (Caughley, 1967) and moufflon (Pfeffer, 1967). It also appears to be common for the males to separate themselves from the females for some time, the extent of bisexual aggregation varying with the species, e.g. in Punjab urial, wild goat, and Asiatic ibex, the males remain with the females throughout the year, whereas in chiru, mountain goat, argali, bighorn and Himalayan tahr, the sexes separate only during certain periods (Schaller, 1977). Moreover, segregation of the sexes also appears to be related to the habitat occupied: in Walia ibex the males spend more time with females than in alpine ibex (Nievergelt, 1974, 1981). Furthermore, according to Couturier (1962) the population density of alpine ibex influences herd composition.

I thank F. Alvarez for revising an earlier draft of this chapter and J. Gonzalez for the computer programming of the statistical analysis.

Material and Methods

Criteria for Age and Sex Classes

Old males: males older than 8 years (those having a black stripe on the flanks, especially conspicuous in winter).

Adult males: males between 4 and 8 years old; their size is bigger than that of adult females and the black stripe on their flanks is thinner than on old males. Their horns, when seen laterally, are S-shaped with the tips pointing upwards and inwards.

Subadult males: males between 2 and 4 years old; same body size as adult females or slightly bigger; horns larger and thicker than adult females. The horns are not yet S-shaped and their tips are bent backwards.

Yearling males: males between 1 and 2 years old; body size smaller than adult females; their horns similar to the adult females, but thicker.

Adult females: horns well developed and maximum body size for females.

Yearling females: females with both horns and body size smaller than adult females. Data concerning this class must be taken with some caution as such females can be confused with adult ones, if seen from afar and if there is no other individual class serving as a reference.

Kids: individuals between 2-3 months and 1 year old; horn size clearly smaller than yearling males or females.

Newly born: animals up to 2 or 3 months old, until their horns are clearly visible.

Group Types

The following group types were considered:

Mixed groups: in which adult or subadult animals of both sexes are found.

Females-kids groups: adult females with kids or newly born. They may be accompanied by yearling males or females.

Yearling groups: yearling males or females, possibly some subadult males as well.

Male groups: males of all ages.

Data Recording

The study was conducted between February 1982 and January 1983.

In July and August few data were collected because of the huge tourist pressure in the study area, making the work difficult and the results less reliable. Thus, data for these months must be taken with some caution and considered suggestive only.

In the study area the ibex live together with four other species of ungulates: red deer, fallow deer, moufflon and wild boar. Hunting and poaching exert a fairly great pressure on all these species. Records included the number of individuals present in each observed group, their age and sex classes, place, habitat, date and time of the day. Observations were made either by car or on foot. When by car, the vehicle was stopped and I waited for the ibex grouped together to flee and, as the flight was not usually fast, it was also possible to record the size of the group and the relevant age and sex classes. When it was not possible to record the full size and composition of the group the observation was removed from the records.

Study Area

The study was conducted in the National Reserve of the Sierras of Cazorla and Segura: an area of 71,000 ha located in the south east of the Iberian Peninsula, latitudes 37°45′ and 38°10′ and longitudes 2°40′ and 3°00′. The highest peak of these mountains is 2,107 m a.s.l.

Areas above 1,200 m have winter snowfalls, the snow remaining all winter over 2,000 m, whereas in the lower areas the snow is scarce and the temperature sometimes reaches above

40°C in summer. The mean annual temperature of these mountains is about 13.5°C. Rainfall varies in different areas of the mountains, generally being around 1,500 mm annually.

The nature of the area includes mostly Mediterranean woodland, ilex being the main shrub, on limestone rocks.

Mountains have been reafforested with *Pinus nigra*, *P. pinaster* and *P. halepensis*, which, together with *Quercus ilex*, are the main tree species.

For the purpose of this study, two habitats with ibex were considered:

a. Open habitat: A low vegetation density, either because of a rocky surface with oaks and other bushes growing in the hollows, or because of grassland with few or no trees. This habitat supplies the ibex with valuable food.
b. Closed habitat: A high vegetation density with space between bushes or trees smaller than 8.6 m.

Results

The percentage of animals present in each of the different kinds of groups observed, referring to the total of animals observed, changes significantly ($\chi^2 = 259.4$, $p < 0.001$ for mixed groups; $\chi^2 = 112.4$, $p < 0.001$ for females with kids groups; $\chi^2 = 21.1$, $p < 0.01$ for yearling groups; $\chi^2 = 130.5$, $p < 0.001$ for male groups) throughout all months of the year (Figure 15.1).

Thus, although mixed groups occur throughout almost all the year except August, the highest ratio of animals in these groups is from November to February (Student-Fisher test, $t = 7.3$, $\alpha = 0.01$). On the other hand, the females-kids groups increase from February onwards and reach their maximum in August, the month in which mixed groups are not found and in which the young, born the previous year, join their mother again after the period dedicated to the birth and care of the newly born (from May onwards) is over. It is worth pointing out that, although the birth season reaches its peak in May, some newly born are still found in the autumn and early winter.

The mixed groups begin to scatter during January and February, forming male groups and females-kids groups. The latter two group sizes increase as the mixed groups decrease,

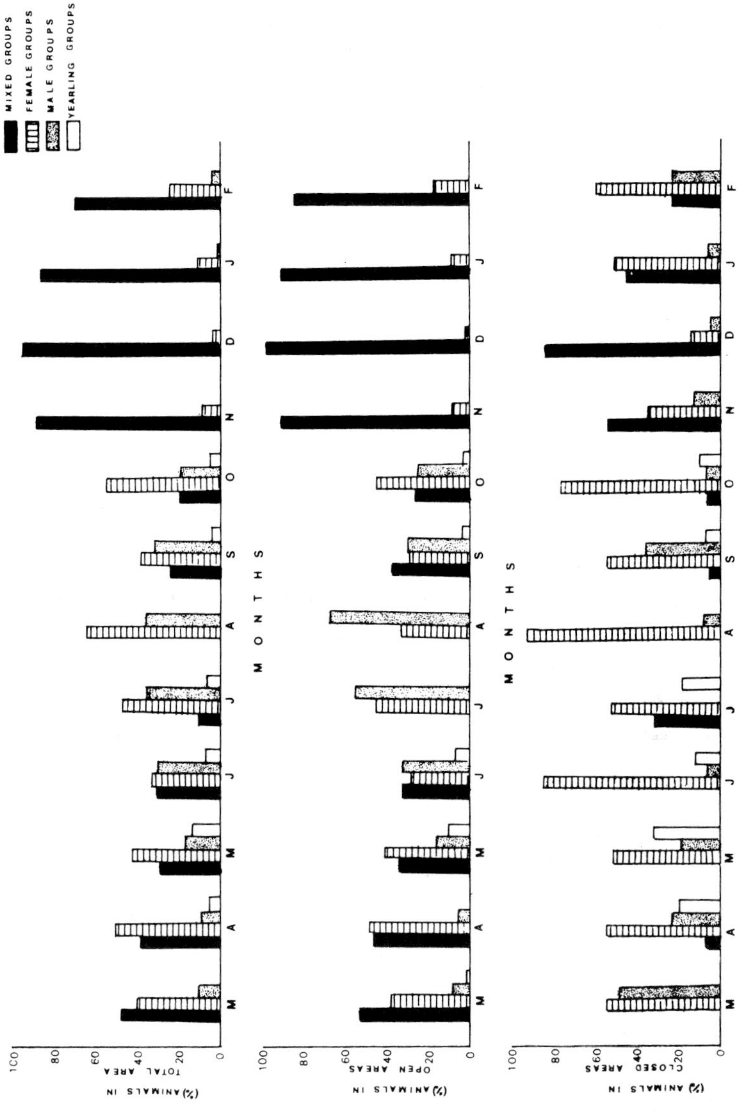

Figure 15.1: Relative Percentages of Animals in Different Group Types Throughout the Year

reaching their maximum in July and August. The male groups reach their peak during the warm months, particularly from June to September (Student-Fisher test, $t = 6.8$, $\alpha = 0.01$).

The frequency of mixed groups is greater in open habitats than in wooded ones throughout the year (Wilcoxon test, $T = 4$, $\alpha = 0.01$), whereas females-kids groups and yearling groups prevail in closed habitats (Wilcoxon test, $T = 4$, $\alpha = 0.01$, $T = 1$, $\alpha = 0.005$ respectively). The male groups are more numerous in wooded habitats than in open ones in February, March, April and May, shifting to open habitats in June, July and August, when the grass begins to grow. Differences are, however, not statistically significant (Wilcoxon test, $T = 39$, NS).

Figure 15.2 (mixed groups) shows that the number of individuals in open habitats exceeds that in wooded ones (Mann-Whitney test, $U = 27$, $\alpha = 0.01$).

The maximum number reached for mixed groups in open habitats coincides with the rut (November-December): sometimes between 40 and 50 individuals at the beginning of the rut. The minimum number occurs in the summer months.

There are more females-kids groups in open habitats (Mann-Whitney test, $U = 24.5$, $\alpha = 0.01$). On the other hand the number of individuals is very uniform in wooded habitats (Kolmogorov-Smirnov test, $D = 0.017$, NS).

A greater number of individuals seems to occur in male groups in open habitats than in wooded ones, reaching its maximum in March in both kinds of habitat, gradually decreasing in the following months, with its minimum at rut time; however, these differences are not significant (Mann-Whitney test, $U = 52.5$, NS).

If one evaluates the distribution of each class in the types of groups and habitats (Figure 15.3), it is observed that most females are found in mixed groups from November to February ($\chi^2 = 235.6$, $p < 0.001$; $t = 6.7$, $\alpha = 0.01$), whereas the majority of them are found in females-kids groups during the remaining months of the year.

In the warm months the percentage of females in wooded habitats increases ($\chi^2 = 186.2$, $p < 0.001$; $t = 5.73$, $\alpha = 0.001$).

The kids are distributed between females-kids groups and mixed groups in the same manner in which their mothers are, except from May to July, during which time some kids can be found in yearling groups (Figure 15.3).

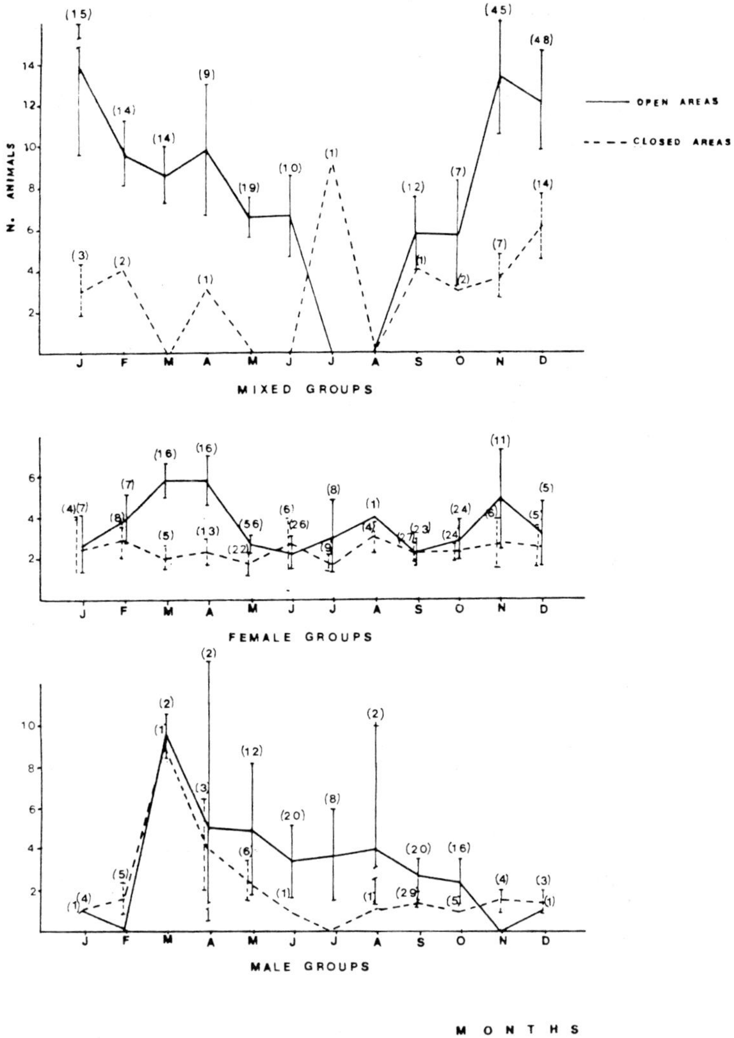

Figure 15.2: Mean Herd Size of the Different Group Types in Open Areas and Closed Areas. The numbers in parentheses show the number of groups observed and vertical lines show the standard error of the mean

The young males are distributed in every type of group. So from November to February (inclusive) they are found mainly in mixed groups and, in the remaining months, in the different groups and habitats (Wilcoxon test, $T = 0$, $\alpha = 0.05$).

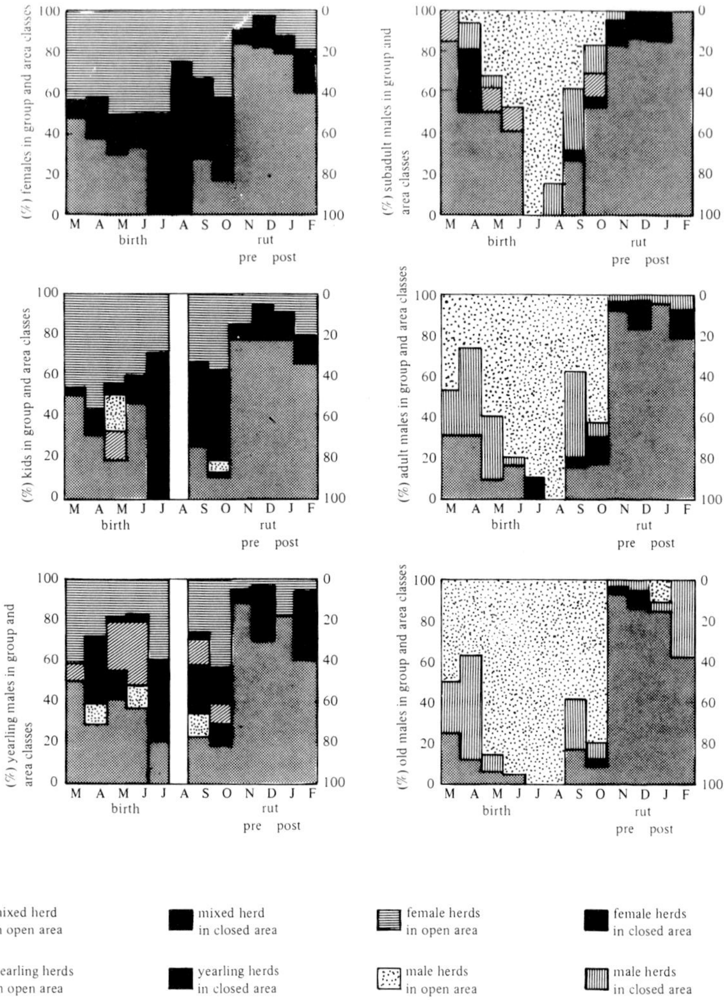

Figure 15.3: Proportion of the Different Types of Groups Located in Open and Dense Vegetation Habitats

The subadult males are found in large numbers in mixed groups from October to May (Wilcoxon test, $T = 0$, $\alpha = 0.05$), although from March onwards they shift to the young groups and male groups, the highest percentage occurring later in the summer.

The shift of subadult males from mixed groups to male groups is gradual, through the yearling groups.

Ibex less than two years old group together from April until October. Their group size reaches its maximum in May, the month when the highest number of births occurs.

The adult males are found in mixed groups in a high proportion (greater than 90 per cent) between the months of November and February, shifting to male groups in the remaining months of the year (Wilcoxon test, $T = 0$, $\alpha = 0.005$). Such a shift seems to occur gradually, when these animals start living in wooded habitats.

Old males spend less time with the females than subadult males ($\chi^2 = 90.1$, $p < 0.001$). So, in November and December, the proportion of old males in the mixed groups is greater than 90 per cent, decreasing in January and February and especially in May.

The movement of old males from some types of groups to others occurs when the animals are in habitats of denser vegetation.

Discussion

If one compares the percentage of individuals living in mixed groups of *Capra pyrenaica* on the Sierras of Cazorla and Segura with the data provided by Nievergelt (1974) for *C. ibex* and *C. walia* (Figure 15.4), the period with the greater proportion of individuals in mixed groups, coinciding with the rut, appears to begin earlier in *C. pyrenaica* than in the other species, and the proportion appears to decrease somewhat earlier. In the remaining months of the year, when the proportion of individuals living in mixed groups is less than the average, the Spanish ibex is generally intermediate between the Alpine and the Walia ibex, probably in relation to relevant intermediate weather conditions.

It is worth emphasising that even outside the rut the Spanish ibex shows a male group size (3.0) intermediate between those recorded for the Alpine and Walia ibex (7.6 and 1.5 respectively).

The separation of males and females to form different groups grazing in separate areas, at higher altitude for male groups, as in *Cervus elaphus* (Burckardt, 1958), is explained, according to

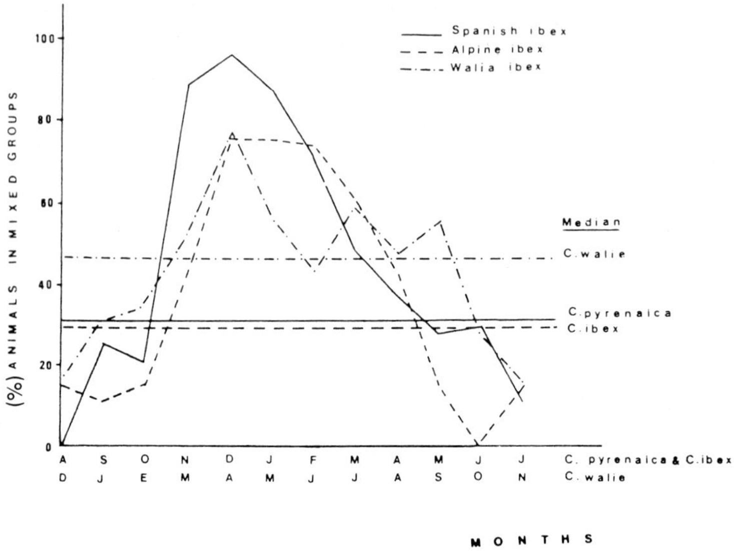

Figure 15.4: Proportion of Mixed Groups in Spanish Ibex, Compared to Walia Ibex and Alpine Ibex (data of the last two species from Nievergelt, 1974).

Geist (1974a,b), by the need of these age classes not to compete for food. This behaviour corresponds to that described for the Alpine ibex (Couturier, 1962): when the population density is high, the adult males and females segregate into separate groups, whereas, when the population density is low, as occurs in recently established colonies, both sexes are grouped together.

The pattern depends on the fact that the yearlings are expelled from the female groups at the time of new birth or immediately afterwards. The first to separate from the female groups are young and subadult males.

Comparable behaviour has been shown for many other species of ungulates (Alpine ibex, Couturier, 1962; feral Soay sheep, Grubb & Jewell, 1966; Himalayan tahr, Caughley, 1967; moufflon, Pfeffer, 1967; feral goats, Rudge, 1970).

The number of individuals living in different kinds of groups changes in accordance with the type of habitat. In habitats with a high density vegetation, the females-kids groups and the yearling groups predominate, probably in relation to the protection that the vegetation provides against predators and man. On the contrary, the mixed groups are found chiefly in habitats with low vegetation density; the larger size of these groups determines the habitat selection.

Male groups are more frequent in wooded habitats during the months in which they are moving from the winter areas to the summer ones, which are mainly open habitats.

In general, for our ibex, group size is bigger in open habitats than in closed ones, just as for other species of ungulates, e.g. roe deer and white-tailed deer (Geist, 1974b), probably because of the advantages that grouping provides against predators (Mirth, 1973; Altmann, 1974; Estes, 1974; Jarman, 1974; Underwood, 1982), or because the higher visibility possibly allows them to maintain the group cohesion (Ewer, 1968; Walther, 1972). In the same way Walther (1978) compared the groups of *Oryx beisa* observed in open habitats with those observed in closed habitats and reached the same conclusions. Jodra (1981) also observed similar behaviour in fallow deer.

The size of mixed groups is greater than that of the other types of groups; this is also true with Punjab urials (Schaller, 1977), fallow deer (Jodra, 1981) and mule deer (Koutnik, 1981), reaching its maximum size on the rut.

The size of female groups in open habitats is somewhat smaller from May to October, contrary to what we might have expected, as in the case of Walia ibex (Nievergelt, 1981). A possible cause of this difference is the effect of tourism on the group size. The grazing areas are the places chosen by tourists to go camping, and this causes the groups to break up and flee to places of greater vegetation density, where it is easier to hide.

In the wooded habitats the group size of females with kids does not change throughout the year. This is not the case for male groups, in which a distinct peak is observed in March, coinciding with the peak in open habitats. The increase in size of the male groups is probably due to the seasonal movement that takes place in March from the winter areas to the spring-summer areas. On the other hand, movement from summer areas to winter areas, in September and October, does not produce large herds, probably because the males do not tolerate one another near the rut.

Older males show a definite peak in staying in mixed groups than younger males, as also occurs in Walia ibex (Nievergelt,

1974, 1981) and Punjab urial (Schaller, 1977). One may conclude that the younger the males, the longer they stay with females (but on the rut, older males chase them out).

Yearling males, when they separate from adult females during the birth and first days of rearing the kids, group together with other yearlings, some even joining the adult male groups, as Couturier (1962) also observed for the alpine ibex.

On the other hand, it is important to emphasise that females and young are hidden in dense vegetation during the warm months, probably because of tourist pressure, which is detrimental to their nourishment as their food requirements would be better satisfied in grassland.

References

Altmann, S. (1974) 'Baboons Space, Time and Energy', *Amer. Zool.*, *14*, 221-48

Alvarez, F., Braza, F. & Norzagaray, A. (1975) 'Estructura social del Gamo (*Dama dama*, Mammalia, Cervidae), en Donana', *Ardeola*, *21*, 89-118

Bertram, B.C.R. (1978) 'Living in Groups: Predators and Prey' in J.B. Krebs & N.B. Davis (eds.) *Behavioural Ecology*, Blackwell, London, pp. 64-9

Brock, V.E. & Riffenbergh, R.H. (1960) 'Fish Schooling: a Possible Factor in Reducing Predation', *J. Cons. Int. Explor. Mer.*, *25*, 307-17

Burckardt, D. (1958) 'Observations sur la Vie Sociale du Cerf (*Cervus elaphus*) au Parc National Suisse', *Mammalia*, *22*, 226-44

Caughley, G. (1967) 'Growth, Stabilization and Decline in New Zealand Populations of the Himalayan Tahr, *Hemitragus jemlahicus*', PhD thesis, University of Canterbury, New Zealand

Couturier, M. (1962) *Le Bouquetin des Alpes*, Arthaud, Grenoble

Cushing, D.H. & Harden Jones, F.R. (1968) 'Why do Fish School?' *Nature, Lond.*, *218*, 918-20

Estes, R. (1974) 'Social Organization of the African Bovidae', in V. Geist & F. Walther (eds.) *Behaviour of Ungulates and its Relation to Management*, IUCN Pub., pp. 166-205

Ewer, R.F. (1968) *Ethology of Mammals*, Logos Press, London

Geist, V. (1974a) 'On the Relationship of Social Evolution and Ecology in Ungulates', *Am. Zool.*, *14*, 205-20

—— (1974b) 'On the Relationship of Ecology and Behaviour in the Evolution of Ungulates: Theoretical Considerations' in V. Geist & F. Walther (eds.) *The Behaviour of Ungulates and its Relation to Management*, IUCN Pub., pp. 235-46

Grubb, P. & Jewell, P.A. (1966) 'Social Grouping and Home Range in Feral Soay Sheep', *Symp. Zool. Soc. Lond.*, *18*, 179–210

Hamilton, W.D. (1971) 'Geometry for the Selfish Herd', *J. Theor. Biol.*, *31*, 295-311

Jarman, P.J. (1974), 'The Social Organization of Antelope in Relation to Their Ecology', *Behaviour*, *48*, 215-67

Jodra, P.J. (1981) 'Paràmetros de gregarismo del Gamo (*Dama dama*) en el coto de Donana', *Doñana Act. Vert.*, *8*, 237-80

Koutnik, D.L. (1981) 'Sex-related Differences in the Seasonality of Agonistic Behaviour in Mule Deer', *J. Mammal.*, *62*, 1-11

Lipetz, V.E. & Bekoff, M. (1982) 'Group Size and Vigilance in Pronghorns', *Z. Tierpsychol.*, *58*, 203-16

Mirth, Ph.D. (1973) *Social Behaviour of White Tailed Deer in Relation to Habitat.* Dissertation Abstr. int., *34*, 1470-1

Nievergelt, B. (1974) 'A Comparison of Rutting Behaviour and Grouping in the Ethiopian and Alpine ibex' in V. Geist & F. Walther (eds.) *The Behaviour of Ungulates and its Relation to Management*, IUCN Publ., pp. 324-40

—— (1981) *Ibex in an African Environment*, Springer-Verlag, Berlin, Heidelberg & New York

Pfeffer, P. (1967) 'Le Mouflon de Corse (*Ovis ammon musimon* Schreber, 1782). Position Systematique, Ecologie et Ethologie Comparées', *Mammalia*, *31* (Suppl), 1-262

Rudge, M.R. (1970) 'Mother and Kid Behaviour in Feral Goats (*Capra hircus* L.)' *Z. Tierpsychol.*, *27*, 687-92

Schaller, G.B. (1977) *Mountain Monarchs*, Wildlife behaviour and ecology series. University of Chicago Press, Chicago and London

Siegfried, W.R. (1980) 'Vigilance and Group Size in Springbok', *Madoqua*, *12*, 151-4

Underwood, R. (1982) 'Vigilance Behaviour in Grazing African Antelopes', *Behaviour*, *79*, 81-108

Vine, I. (1971) 'Risk of Visual Detection and Pursuit by a Predator and the Selective Advantage of Flocking Behaviour', *J. Theor. Biol.*, *30*, 405-22

Walther, F. (1972) 'Social Grouping in Grant's Gazelle (*Gazella granti* Broke, 1872) in the Serengeti National Park', *Z. Tierpsychol.*, *31*, 348-403

—— (1978) 'Behavioural Observations on Oryx Antelope (*Oryx beisa*) Invading Serengeti National Park, Tanzania,' *J. Mammal.*, *59*, 243-60

CHAPTER 16

Seasonal Changes in the Size of Chamois Groups in the Ammergauer Mountains, Bavaria

IRMGARD VON ELSNER-SCHACK
Institute of Wildlife Research, University of Munich, Amalienstrasse 52, D-8000 München 40 (FRG)

Abstract

Chamois group sizes were studied in a population inhabiting a Bavarian mountain range. In general, the chamois group sizes were highly variable at all seasons of the year. Single animals were observed frequently. The largest groups occurred during the snow-free periods of late summer and autumn. It is suggested that variation in group size is related to the distribution of food resources.

Introduction

This paper presents information on the variation in the group sizes of chamois (*Rupicapra rupicapra*) which occurs in response to season, altitude and vegetation.

It was hypothesised that group size is related to the fact that plentiful food is found only in patches. This hypothesis appears to hold for winter, spring and autumn but has to be rejected for summer.

Methods and Study Area

The study area is in the Ammergauer Mountains, north of Garmisch, in Bavaria. A description of the study area is given in

v. Elsner (1980, 1982). During one year of field work in 1978/79, 3,725 chamois were observed in 1,395 groups consisting of 1 to 39 animals. Surveys were carried out every three weeks on standardised routes. Three to five surveys were pooled to form the time periods used in the various analyses.

It was observed that groups broke up and regrouped continuously. A 'group' in this paper therefore refers to instantaneous spatial aggregations of chamois rather than long-term associations of individuals. To analyse the effects of seasonality on group sizes, presence of snow, altitude and vegetation were examined. The study area was divided into five equal parts. Type of vegetation where each group was found was determined by reference to a vegetation map with a 4 ha grid. All statistical tests were χ^2.

Results

Most papers report on mean group sizes. Means are appropriate in describing tendencies in normal distributions. In highly skew distributions, such as in group size of ungulates, means provide little biological information. They place too much importance on large groups. A more realistic picture is obtained by using a median which divides the population into two halves of equal size, each containing 50 per cent of all chamois observed. In this case values for medians will be smaller than for means.

Figure 16.1 shows the distribution of group sizes during winter (5 surveys between 4 December and 26 March, 889 chamois in 325 groups) and summer (5 surveys between 10 July and 20 February, 1464 chamois in 469 groups). Group sizes ranged from one to 20 in winter and one to 39 in summer. Medians are 4.0 and 4.9. Statistical analysis indicated a 30 per cent certainty that seasonal distribution was different.

Seasonal variations in group size were largely due to the presence of snow. Consequently the surveys were grouped according to whether snow was present or not. Median group size varied between 3.8 (N=559 chamois) for snow covered periods, 2.7 (N=1,177) during partly covered periods and 4.9 (N=1,685) during snow free times.

Two group sizes were then analysed; single chamois and groups of more than 10 animals. There was a significant

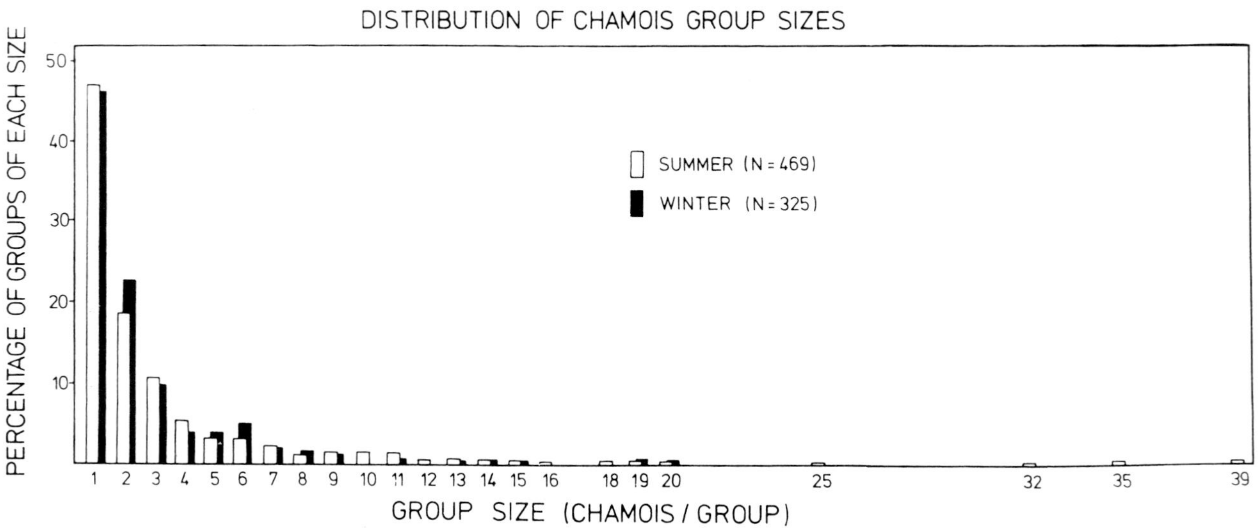

Figure 16.1: Percentage of Different Size Groups of Chamois in Winter (black columns) and Summer (white columns). The median for group sizes changes from summer to winter. Maximum group sizes occur during summer only.

Table 16.1: Use of Altitude by Numbers of Large Groups over Time (≥ 11 Chamois)

Altitude (m)	Season					
	6 March-7 May	8 May-9 July	10 July-10 Sept.	11 Sept.-12 Nov.	13 Nov.-14 Jan.	15 Jan.-25 Feb.
1,591-1,880			3	1		
1,411-1,590			2	2	1	
1,181-1,410			1		1	
966-1,180	1				1	
820-965						

Table 16.2: Use of Vegetation by Large Groups (≥ 11 Chamois)

Altitude (m)	Season					
	6 March-7 May	8 May-9 July	10 July-10 Sept.	11 Sept.-12 Nov.	13 Nov.-14 Jan	15 Jan.-25 Feb.
1,591-1,880			grassland/ krummholz	grassland		
1,411-1,590			grassland/scree	grassland	grassland	
1,181-1,410			scree		spruce/pine	
966-1,180	mixed forest					
820-965						

difference in the distribution of chamois at snow free periods as compared with snow covered times ($p < 0.1$, N = 120, N = 244). They did not use the lowest altitudes during snow free times. A similar analysis for large groups was not carried out due to shortage of data. These large groups were observed mainly at altitudes above 1,400 m, less frequently at altitudes between 1,000 m and 1,400 m and not at all below 1,000 m (Table 16.1).

Single animals used a variety of vegetation types. Table 16.2 indicates that large groups used open vegetation types more than forest; in fact only open stands of forest were used. This might be due to poor visibility in dense forest cover as compared to open stands of forest or open vegetation types. For members of a group, visibility diminishes in forest cover and might prevent chamois from organising into large groups.

Discussion

I suggest that distribution of food resources is one of the main factors determining group size in my study area. During winter, exploitable feeding areas are few and small. In this coarse-grained habitat, chamois are forced to feed together in fairly large groups (median = 3.8).

During spring and autumn there is a patchwork of scattered small feeding areas. In this fine-grained habitat, chamois must disperse into small scattered groups (median = 2.7). By contrast, during the summer, vegetation is almost uniformly of high quality and easily accessible. In fact, the habitat is one large patch. This does not influence group size and the large groups (median = 4.9) found in summer must be explained by another factor.

Krämer (1969) suggests that chamois have a ‘need’ to join social units. The above explanation shows other reasons for grouping. The lack of these reasons for summer grouping could mean that there are overriding factors in operation, such as socialisation of newborn kids and higher security in large groups where alertness is increased. There must be more than strictly social or environmental factors to influence group size of chamois. It would be useful to study different population densities within the same habitat to find out if distribution of chamois is very density dependent.

References

Elsner, I. von (1980) 'Zur Raumnutzung von Gams (*Rupicapra rupicapra* L.) — Untersuchungen in den Ammergauer Bergen', *Diss. Univ. Heidelberg*, pp. 109

—— (1982) Überleben im Hochgebirge: Eine Studie zur Raumnutzung von Gams', *Zeitschr. f. Jagdwiss. (28)*1, 18-30

Krämer, A. (1969) 'Soziale Organisation und Sozialverhalten einer Gemspopulation (*Rupicapra rupicapra L.*) der Alpen', *Z. f. Tierpsychol.*, *26*, 889-964

CHAPTER 17

Fecundity Patterns of Stable and Colonising Chamois Populations of New Zealand and Europe

J.J. BAUER
Dorfstrasse 46, 7830 Emmendingen (FRG)

Abstract

Sexual maturation and fecundity of stable and colonising populations of chamois (*Rupicapra rupicapra* L.) in New Zealand and Europe are compared. It is evident that early sexual maturation at the age of 1.5 years or even seven months is not only a feature of colonising populations (dispersal front), but is also found within stable populations (central distribution), which are living under suboptimum growth conditions and show reduced body growth. This pattern contrasts strongly with European populations, where sexual maturation at the age of 1.5 years seems to be confined to colonising populations. If fecundity patterns are compared with growth patterns, this discrepancy can be demonstrated clearly. Whereas in Europe early maturation of female yearlings in colonising populations is correlated with corresponding rapid juvenile weight gain, as would be expected, the two New Zealand populations compared do not show this pattern. Even under sub-optimum growth conditions, which are realised to their extreme in the Harper-Avoca region, a substantial amount of females reach maturation at the age of 18 months, although body weights are below 14 kg (eviscerated). Whereas a reduction in size of 18-month-old females within European stable populations of 15 per cent is accompanied by a fecundity reduction of more than 90 per cent, corresponding values within New Zealand populations lie around 40 per cent and 35 per cent. The implications of this change in fecundity pattern are discussed.

Introduction

The age of sexual maturation in chamois is subject to large variation. In stable populations of the European Alps chamois start reproducing at the age of four to even five years (Couturier, 1938; Krämer, 1969; Schröder, 1971). In colonising populations, established from recently introduced animals, like in the Swiss Jura (Salzmann, 1977), the Black Forest (Gossow, 1978) and the Vosges (Gossow, 1978; Pflieger, 1978) it is known, or at least suspected, that a certain amount of 2-year-old females give birth. In New Zealand, sexual maturation at the age of 1.5 years has been recorded in all populations investigated so far (Caughley, 1970b; Lambert & Bathgate, 1977; Bauer & Gossow, 1978; Bauer, 1982).

During earlier studies it was established that the onset of sexual maturation of chamois (males and females) of specific populations is the function of a threshold body weight reached during a specific daylength (Bauer, 1982). The aim of this study was to compare fecundity patterns of populations existing under different environmental conditions and to try to elucidate links between growth, size and fecundity.

I wish to thank R. Henderson, L. Batchelor, C. Clarke, H. Gossow, K. Westerskov, A. Duncan, D. Klimetzek and M. Crawley as well as the German Academic Exchange Service (DAAD), the New Zealand Forest Service, The University of Otago and the University of Canterbury for aid during various stages of the study.

Material and Methods

During a study on the dynamics of New Zealand chamois populations, data on reproduction and physical condition of 90 females from populations in the southwest of the South Island (Lake Wanaka) and from 54 females from populations in the Central Alps (Harper-Avoca), west of Christchurch, were collected (Figure 17.1). The fecundity values (m_x = number of female offspring per female of a given age) obtained during this study as well as data on 326 chamois females from the Rangitata/Copland region collected by Caughley (1970b) and 101 females from the Arawata region (Lambert & Bathgate, 1977)

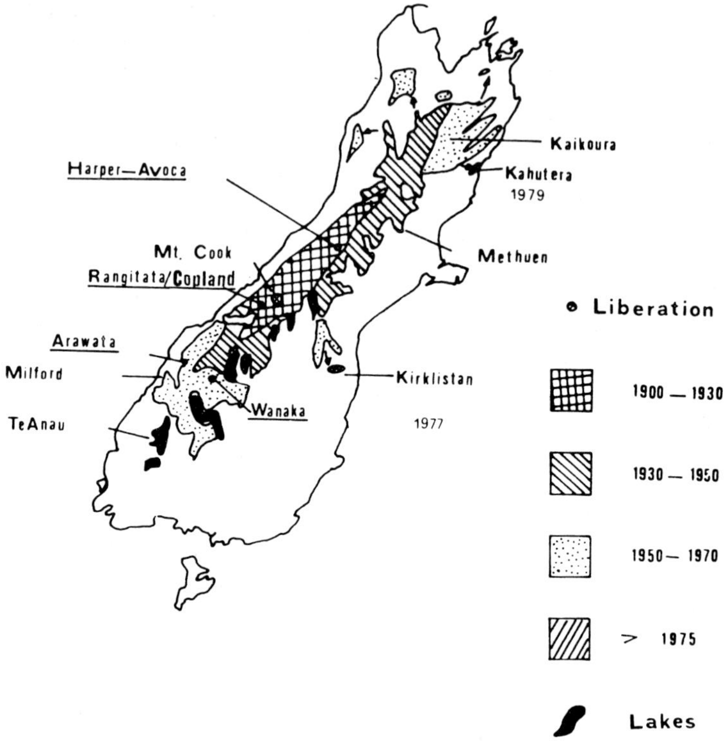

Figure 17.1: Dispersal of Chamois in New Zealand. Liberation site: Mt. Cook. Areas mentioned in the text are underlined (modified map of the New Zealand Forestry Service)

have been chosen to represent reproductive performance of New Zealand chamois.

The four populations compared fall into two distinct groups. The Rangitata/Copland and Harper-Avoca regions (Figure 17.1) were colonised by chamois between 1915 and 1925 (Caughley, 1970b; Logan, 1956) and are in a posteruptive stage (Caughley, 1970a) and stable in numbers (Caughley, 1970b; Clarke & Henderson, 1978), therefore I will term them stable. The Wanaka and Arawata regions (Figure 17.1) have been colonised between 1945 and 1960 (Lambert & Bathgate, 1977) and the chamois are still in the process of moving into adjacent previously unoccupied areas. These populations will be classified as colonising. Furthermore typical conditions for colonising populations (i.e. overabundance of food) or stable populations (depletion of food resources) were enhanced by the heavy exploitation by helicopter hunting of the Wanaka and Arawata regions, whereas Rangitata/Copland and Harper-Avoca populations were comparatively little or not affected by this matter at the time of sampling. These differences are well demonstrated by physical characteristics like body growth and body size (Bauer, 1982). Body weights were only available for the Harper-Avoca and the Wanaka populations.

Fecundity values of European populations are sparse and, if available, derived from small samples. However, data could be used for three alpine populations, the Hochschwab (Schröder, 1971), the Augstmatthorn region (Krämer, 1969) and the Appenzell/Schwyz region (Hofmann, 1978). These populations are called stable, although the application of this criterion is somewhat questionable for the last one and must be taken with caution. Two populations derived from a few recently (1937-1957) introduced animals in the Swiss Jura (Salzmann, 1977) and the Black Forest (Gossow, 1978) fulfil the criterion 'colonising'. Data on body weights of European chamois were available for three stable populations from the Hochschwab (Schröder, 1971), the Graubünder region (Ratti, 1975) and the Appenzell/Schwyz (Hofmann, 1978), two regions where chamois are colonising, and from the Swiss Jura (Salzmann, 1977) and the Vosges (Pflieger, 1978).

Results

Fecundity Tables of New Zealand Chamois

Sexual Maturation and Fecundity in Colonising Populations. Although reproductive performance of chamois of both regions is comparable, Arawata females show considerably lower overall productivity. Whereas Wanaka yearling females had all undergone oestrus or were carrying a fetus, less than a third of Arawata yearling females gave birth (Table 17.1). In the Wanaka region even kid pregnancy seemed to be no exception.

Sexual Maturation and Fecundity in Stable Populations. The fecundity values of chamois females of both stable populations are very similar. More than one-third of the females had given

Table 17.1: Age-specific Fecundity Values (m_x) of Two New Zealand Chamois Populations in Recently (< 30 years) Colonised Areas (Arawata figures from Lambert & Bathgate, 1977)

Age (years)	Wanaka N	Wanaka m_x	Arawata N	Arawata m_x
1	11	0.091	19	0.000
2	10	0.500	11	0.136
3			20	0.325
4			17	0.411
5			7	0.357
6	30	0.500		
7				
8			9	0.389
9				
10 +	6	0.333		
Total	57		83	

Table 17.2: Age-specific Fecundity Values (m_x) of Two New Zealand Chamois Populations in Areas Colonised for More than 50 Years (Rangitata/Copland figures from Caughley,1970b)

Age (years)	Rangitata/Copland N	Rangitata/Copland m_x	Harper-Avoca N	Harper-Avoca m_x
1	60	0.016	12	0.000
2	36	0.194	7	0.214
3	70	0.371	3	0.166
4	48	0.469		
5	26	0.365		
6	19	0.421		
7	6	0.417	24	0.452
8	2			
9	3	0.350		
10 +	5		7	0.429
Total	275		53	

birth already at the age of two. In the Rangitata/Copland area, a small number of kids were sexually mature (Table 17.2), a phenomenon also encountered in the colonising Wanaka population (Table 17.1).

Sexual Maturation and Growth Patterns

As sexual maturation in mammals (Sadlier, 1969; Amir, Kali & Volcani, 1968) and in this case in chamois (Bauer, in preparation) is a function of physical development, which in turn is determined by the availability and quality of nutrition, the question arises if this lack of difference is due to non-existing differences in growth of stable and colonising chamois populations. In other words, the concept of 'stable' or 'colonising' populations may not be valid! This however is not the case. The criteria 'stable' and 'colonising' in New Zealand chamois populations can clearly be qualified by large differences in body growth between local populations due to food depletion.

Comparing body weight development of New Zealand and European chamois populations demonstrates these differences (Figure 17.2). In New Zealand as well as in Europe growth patterns of subadults of stable populations differ from those of colonising populations (Figure 17.2). Whereas 18-month-old females of European stable populations (N=3) show an average body weight of 15.4 kg, 18-month-old females of colonising populations (N = 3) weigh 18.3 kg.

Although only two New Zealand populations could be compared, the difference is even more pronounced (Figure 17.2, Table 17.3). This indicates that differences in growth conditions between New Zealand chamois populations show higher extremes than within European populations so far investigated.

If these growth patterns are compared with fecundity patterns, discrepancies between New Zealand and European chamois populations emerge, which are difficult to interpret.

Fecundity Patterns of New Zealand and European Chamois Populations

In Figure 17.3 fecundity patterns of New Zealand and European chamois populations (N=9) are compared. Here, according to more pronounced differences in juvenile development within New Zealand chamois (Table 17.3) it should be expected that within the first reproducing age class more pronounced differ-

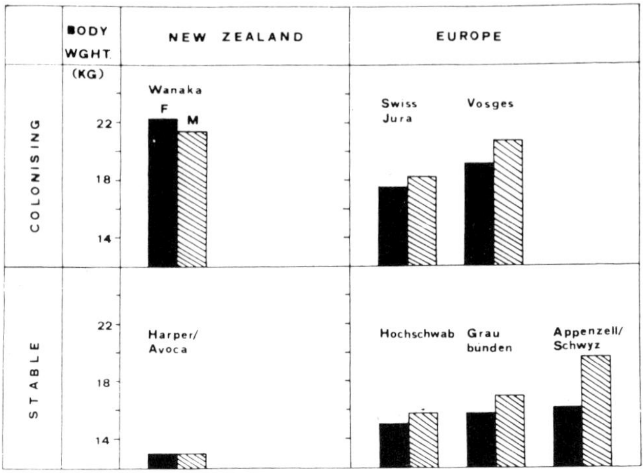

Figure 17.2: Average Body Weights of 18-month-old Chamois Males (M) and Females (F) in Four Stable and Three Colonising Populations of New Zealand and Europe (Data from: Salzmann, 1977; Pflieger, 1978; Schröder, 1971; Ratti, 1975; Hofmann, 1978; Bauer, 1982, 1983a)

Table 17.3: Average Body Weights of 13-month-old Chamois Females of Stable and Colonising Populations of New Zealand and Europe

	New Zealand		Europe	
	$\bar{x} \pm s$	N	$\bar{x} \pm s$	N
Stable	13.3	1	15.35 ± 0.49	3
Colonising	22.3	1	18.3 ± 1.13	3

Table 17.4: Fecundity Values (m_x) of 18-month-old Chamois Females of Stable and Colonising Populations of New Zealand and European Chamois

	New Zealand		Europe	
	$\bar{x} \pm s$	N	$\bar{x} \pm s$	N
Stable	0.204 ± 0.014	2	0.017 ± 0.029	3
Colonising	0.318 ± 0.257	2	0.251 ± 0.117	2

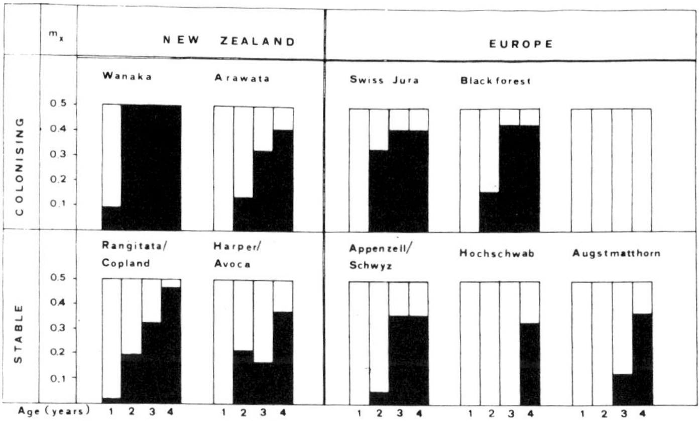

Figure 17.3: Fecundity Values (m_x) of 1- to 4-year-old Chamois Females of Four Colonising and Five Stable Populations of Chamois in New Zealand and Europe (Data from: Lambert & Bathgate, 1977; Salzmann, 1977; Caughley, 1970b; Gossow, 1978; Hofmann, 1978; Schröder, 1971; Krämer, 1969; Bauer, 1982, 1983a)

ences in fecundity values do exist. This however is not the case. Differences in the fecundity of yearlings between two stable and two colonising populations are inconclusive. All the four populations show relatively high fecundity values of yearling females (Figure 17.3, Table 17.4). In stable as well as in colonising populations (Tables 17.1 and 17.2) sexually mature 7-month-old kids do occur.

If reduction in body weight and corresponding reduction in fecundity in chamois populations of New Zealand and Europe after their stabilisation are compared, this difference is evident (Figure 17.4). Whereas in Europe a reduction in physical juvenile development of 16 per cent is accompanied by a reduction in fecundity of 93.2 per cent, a body weight reduction in 18-month-old New Zealand chamois females of stable populations (in relation to colonising populations) of 40.4 per cent leads to a fecundity reduction of only 35.8 per cent (Figure 17.4).

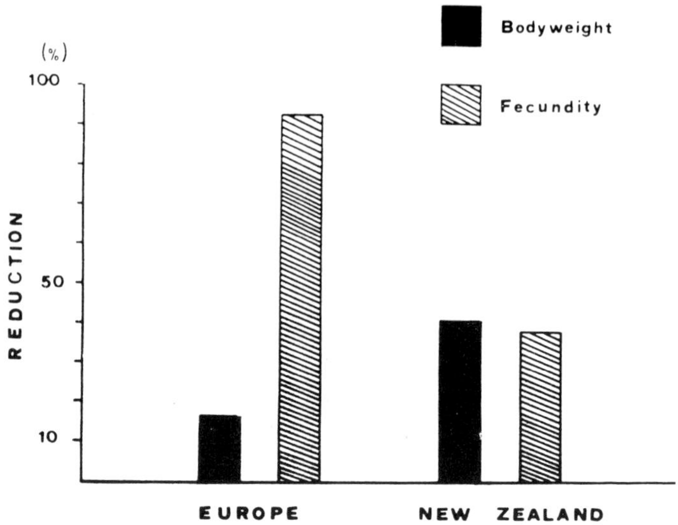

Figure 17.4: Relative Reduction of Body Weight and Fecundity of 18-month-old Chamois Females in Stable Populations (in relation to colonising populations) in Europe and New Zealand

Discussion

The rate of increase of animal populations, according to Cole (1954) and Lewontin (1965), is very sensitive to changes in the age at first reproduction. In increasing populations, with abundance of food, sexual maturation will be pushed down to the physiological limit of the species.

In natural populations, however, growth rate fluctuates around zero and only rarely will a population experience a long uninterrupted period of increase or decline. Thus, pure *r*-selection, or pure selection for delayed maturation, represent hypothetical stages seldom occurring in nature (Stearns, 1976). Species-specific variations of the age of sexual maturation are therefore consequences of selective forces opposing each other at different population stages (increase or decline).

According to models of island populations, proposed by MacArthur (1967), a colonising species multiplies rapidly in a new suitable environment, then declines slowly. Continuing selection pressure for early reproduction during the colonisation phase (Pianka, 1970) might lead to less genetic variation of this specific life-history trait. Thus, establishment of low age of maturation might result, although declining food resources and population stabilisation in accordance with the carrying capacity of the environment should have favoured delayed maturation again.

It seems likely that during the expansion of New Zealand chamois, which led to the colonisation of nearly 40,000 km^2 in the course of seventy years, in a country formerly unoccupied by mammals, the chamois found conditions comparable to those existing during the late Pleistocene in Europe.

According to Geist (1971), 'it appears that after an initial size increase during colonisation the individuals of a species decline in size . . . the reduction in size could be due to selection for longevity in qualitatively poor populations and consequently the evolution of slower growing, small individuals.' Geist (1971) substantiates this hypothesis by palaeontological evidence from the Villafranchian, the period in earth history when large mammals like aurochs (*Bos*), goral (*Nemorhaedus*), tahr (*Hemitragus*), first sheep (*Megalovis*), true goats (*Soergelia*) and many other species, including man, moved into Europe and Asia. All of them declined in size, after seemingly having reached maximum size.

New Zealand chamois show this pattern of growth on the population level. During the early colonisation phase, after reaching dimensions unequalled in Europe (Bauer, 1977), final body size of chamois population above or at carrying capacity is considerably smaller than in Europe. This applies not only for the Harper-Avoca region, but seems to be the case in the central portion of the New Zealand Alps (Bauer, 1982).

Further evidence supports this hypothesis. Decreased sexual dimorphism of New Zealand's expanding chamois populations at their periphery of dispersion could be shown to be considerably smaller than within European or stabilised New Zealand populations (Bauer, 1977; Bauer & Gossow, 1978). This is in accordance with Geist's (1971) 'Neotenisation Hypothesis', where expanding populations of mammals, including man, show reduced sexual dimorphism.

Although Geist's theory might explain physical characteristics of stable and colonising New Zealand chamois populations, other causes have to be sought to explain the observed fecundity patterns. These causes might be found in specific environmental conditions encountered in the New Zealand Alps, e.g. the extension of the growth period of chamois because of a longer period of positive energy balance. It seems unlikely that this major change in environmental conditions should have had no consequences on the growth rate of this species. It seems possible that extension of the growth season might lead to early maturation, despite bad growth conditions, small adult body size in turn being determined by the termination of growth because of sexual maturation (Carlson, 1969). Another explanation might be found in the unusual mortality patterns of the Harper-Avoca population with a high mortality of yearlings and two-year-old animals (Bauer, 1982; Bauer & Gossow, 1983).

References

Amir, S., Kali, J. & Volcani, A. (1968) 'Influence of Growth Rate on Reproduction and Lactation in Dairy Cattle' in Lodge, G.A. & Lamming, G.E. (eds.) *Growth and Development of Mammals*, Butterworths, London, pp. 234-6

Bauer, J.J. (1977) Population Dynamics of Chamois at Lake Wanaka, New Zealand, Thesis, University of Otago, Dunedin, N.Z. 78 pp

—— (1982) *Untersuchungen zur Dynamik von Stabilen und Kolonisierenden Gemsenpopulationen (Rupicapra rupicapra L.) in Neuseeland*. Inaug. Diss. Biol. Fak. Univ. Freiburg i. Brsg. 201 pp

—— (1983a) 'Selection for Low Age of Maturity in New Zealand Chamois (*Rupicapra rupicapra* L.), *15th Pac. Sci. Congr.*, Dunedin, NZ, Abstr. p. 15

—— (1983b) 'Beobachtungen zur Biologie und Problematik neuseeländischer Gemsenpopulationen', *Jagd u. Hege*, *15*, 9-11

Bauer, J.J. & Gossow, H. (1978) 'Zur Populationsdynamik der Gemse in Neuseeland'. *3 Int. Gamsw. Symp. Mayrhofen*, Tirol, pp. 63-74

Bauer, J.J. & Gossow, H. (1983) 'On Diseases as a Major Regulatory Factor in the Population Dynamics of N.Z. Chamois', *15th Int. Congr. Game Biol.*, CSSR. Abstr.

Carlson J.R. (1969) 'Growth Regulators' in Hafez, E.S.E. and Dyer, I.A. (eds.), *Animal Growth and Nutrition*. Lea & Febiger, Philadelphia, pp. 138-55

Caughley, G. (1970a) 'Eruption of Ungulate Populations, with Emphasis on Himalayan Thar in New Zealand', *Ecology*, *51*, 53-72

—— (1970b) 'Population Statistics of Chamois', *Mammalia*, *34*, 194-9

Clarke, C.M.H. & Henderson, R. (1978) 'Snare Marking Chamois in New Zealand', *Z. Säugetierk.*, *43*, 232-9

Cole, L.C. (1954) 'The Population Consequences of Life-history Phenomena', *Quart. Rev. Biol.*, *29*, 103-37

Couturier, M.A.J. (1938) *Le Chamois*. Arthoud, Grenoble

Geist, V. (1971) *Mountain Sheep — a Study in Behaviour and Evolution*, The University of Chicago Press, Chicago and London

Gossow, H. (1978) 'Krukenwachstumsmuster als Weiser in der Gamswild-bewirtschaftung', *3rd Int. Gamswild Symp.*, Mayrhofen, Tirol, pp. 206-19

Hofmann, H. (1978) *Unterschiedliche Entwicklungsleistungen von Appenzeller und Schwyzer Gemsen (Rupicapra rupicapra L.) und ihre möglichen Ursachen*. Inaug. Diss. Univ. Zürich, Schweiz

Krämer, A. (1969) 'Soziale Organisation und Sozialverbalten einer Gamspopulation (*Rupicapra rupicapra* L.) der Alpen', *Z. Tierpsychol.*, *26*, 889-964

Lambert, R.E. & Bathgate, J.L. (1977) 'Determination of the Plane of Nutrition of Chamois', *Proc. N.Z. Ecol. Soc.*, *24*, 48-56

Lewontin, R.C. (1965) 'Selection for Colonizing Ability' in Baker, H.G. & Stebbins, G.L. (eds.), *The Genetics of Colonizing Species*, Academic Press, New York and London, pp. 79-92

Logan, P.C. (1956) 'Some Problems in the Control of Wild and Domestic Animals in a Critical Hydroelectric Catchment', *N.Z. J. For.* *7*: 68-75

MacArthur, R.H. (1967) *The Theory of Island Biogeography*, Princeton University Press, Princeton

Pflieger, R. (1978) 'Gewicht und Schädelmaße der Vogesengams', *3rd Int. Gramswild Symp.*, Mayrhofen, Tirol. pp. 167-81

Pianka, E.A. (1970) 'On r and K-selection', *Amer. Natur.*, *100*, 592-7

Ratti, P. (1975) 'Bericht über Untersuchungen der Bündner Gamsstrecken der Jahre 1971-1974', *Bündner Jägerzeitung*, *62*, 89-95

Sadlier, R.M.F. (1969) *The Ecology of Reproduction in Wild and Domestic Mammals*, Methuen & Company, London, 321 pp

Salzmann, H.C. (1977) 'Untersuchungen zur Fortpflanzungsbiologie der Gemsen des schweizerischen Jura', *Z. Säugetierk.*, *42*, 180-9

Schröder, W. (1971) 'Untersuchungen zur Ökologie des Gamswildes in einem Vorkommen der Alpen', *Z. Jagdwiss.*, *17*, 113-68, 177-235

Stearns, S.G. (1976) 'Life History Tactics: A Review of the Ideas', *Quart. Rev. Biol.*, *51*, 3-47

CHAPTER 18

Social Grouping Patterns of a Dense Population of Chamois in the Western Pyrenees National Park, France

C. BERDUCOU[1] and P. BOUSSES[2]

1. Centre d'Etude et Recherche sur la Faune de Montagne, 64880 Etsaut (France)
2. Laboratoire de Neuroethologie, Université de Toulouse, 117 route de Narbonne, 31000 Toulouse (France)

Abstract

A very dense population of Pyrenean chamois (*Rupicapra rupicapra pyrenaica*) occurs in the Pic du Midi d'Ossau area, inside the Pyrenees National Park.

Very large females-kids flocks (up to 208 individuals), large summer groups of adult males (up to 64 individuals) and continuous high rate of mixing of the individuals into the different groups of large 'social units' are the most characteristic social features of this population. A standard method of studying and representing social grouping patterns is proposed.

Introduction

The populations of chamois living on the French side of the Pyrenees are managed so that there are very low densities in the hunted areas as opposed to very dense nuclei in the game preserves and in the National Park. We studied the nucleus which is the most advanced in the process of demographic expansion. Our aim was to devise a standard method of recording and representing social grouping patterns which is simple enough to be currently used by management services, thus providing data for effective comparisons between populations, and to detect changes in the evolution of a population. The hypothesis to be tested in further research is that social grouping patterns, being highly density-dependent, could provide an early warning sign of demographic expansion beyond the carrying capacity of the habitat.

Material and Methods

Three methods were used to record field data: the 'Flash-pointing', the 'Periodic itinerary', and the 'Fixed panoramic observation site'.

The 'Flash-pointing' census consists of a rapid and simultaneous screening of the whole study area by 15 teams of Park wardens (Berducou, Besson & Gardes-Moniteurs du PNPO, 1982). Chamois locations were plotted on special 1:25,000 maps and their groups were described on standard forms. This record was carried out twice a year: first at the end of the birth period (mid-June), and secondly, in the rutting season (mid-November).

The 'Periodic itinerary' consists of a similar method applied to a fairly small sampling area inspected on foot twice a week during summer, and at least once per month in the other seasons.

The 'Fixed panoramic observation site' method consists of a continuous survey of a wide area, carried out from sunrise to sunset.

A 'Scan sampling' (Altmann, 1974) was done every 30 minutes, and chamois locations were plotted on a large photograph; group size, group composition, and individual activities were recorded on special forms.

In all these methods, 20 × 80 binoculars were used to classify each individual into one of the four basic age-sex categories (Krämer, 1969): chamois on their first calendar year (1st Y), chamois in their second calendar year (2nd Y), adult males (♂♂), and adult females (♀♀).

The term 'group' was used in the sense of several animals staying less than 40 m apart from the others at the time of the observation (Krämer, 1969). An isolated chamois was considered as a 'group' of one animal.

A few chamois were individually marked with brightly coloured plastic sheaths around the horns, after being caught by foot-snares (Berducou & Apollinaire, in preparation).

Study Area

The Pic du Midi d'Ossau mountain area is located on the western part of the Pyrénées National Park, along the Spanish border. The planimetric extension of the area is 8,800 ha. The altitude ranges from 1,093 to 2,884 m a.s.l. The fauna has not been hunted since 1949. The population dynamics of these chamois has been described by Berducou *et al.* (1982). Their numbers reached a maximum of 1,685 in 1977, then they slowly decreased to 1,235 by 1981, when this study began. At that time, mean density was 14 chamois/100 ha in the most favourable sites. Birth rate was 0.69 kids/adult female, and sex-ratio in the adult class was 1:2. Overbrowsing of forest was evident in the winter ranges.

Results and Discussion

Group Size

Four 'Flash-pointing' censuses were carried out. The results of those conducted in November 1980 and November 1981 were pooled together (2,153 chamois of which 1,846 were age-sex identified, in 262 groups); the same procedure was used in the censuses conducted in June 1981 and June 1982 (2,808 chamois, of which 2,289 were sexed and aged in 240 groups).

Table 18.1 shows how the chamois were distributed in groups of different sizes. In June, over half of the chamois were in groups of more than 25 and one-third were in groups of more than 50. Mean group size was 11.7, and the maximum number observed in a single group was 208 individuals. In November more than one-third of the chamois were in groups larger than 25. The mean group size was 8.2, and the maximum observed was 82.

By limiting the statistics to the groups containing all the four age-sex categories, we found that mean group size was 58.8 in June and 22.1 in November. In similar density conditions, Krämer (1969) observed flock sizes of 25.1 in July-August and 21.5 in November-December, with a maximum of 25.2 in September-October. Schröder (1971) observed flock sizes of 12.4 in June and 17.3 in November, with a maximum of 36.7 in August. Lovari (in press) reported a mean group size (females and subadults) of 16.7 in November.

Table 18.1: Distribution of the Chamois of the Ossau Mountain Area in Social Groups of Different Size Classes. Percentage of chamois in parentheses

Months	Group size classes						
	1	2	3-10	11-25	26-50	51-100	>100
June	(2.9)	(2.8)	(17.0)	(21.0)	(22.8)	(17.2)	(16.3)
November	(3.1)	(3.0)	(29.8)	(29.6)	(25.5)	(9.0)	—

Although our data are only relevant to the months of June and November, we could assess that the maximum aggregation of chamois took place immediately after the birth period. The occurrence of very large groups, especially in June, was the most striking feature of the Ossau chamois population.

Sex-age Class Grouping Patterns

In order to express how the different age-sex classes mix together in social groups, we have selected a standard easy-to-read diagram which can provide a direct visual comparison between different censuses. For that purpose, we started from a triangular diagram (see Figure 18.1A) in which each group is plotted so that its coordinates represent the relative ratios of males (top), females (left) and juveniles (1st Y and 2nd Y, right), e.g. group G in Figure 18.1A includes 20 per cent males, 30 per cent females and 50 per cent juveniles. By separating the different lots of points, we traced the standard diagram shown in Figure 18.1B, in the cells of which one can indicate the numerical values or signs relevant to the corresponding social group type.

From the data recorded by 'flash pointing' censuses, Figure 18.1C shows the relative distribution of the chamois in the different group types. Figure 18.1D, and Figures 18.2E, F and G deal respectively with the distribution of males, females, 1st Y and 2nd Y chamois. Comparison can be made easily at the first glance by this semi-graphic arrangement.

The most striking result is that adult males were seldom alone. In June, the majority of the males were found in unisexual groups of five individuals (mean size).

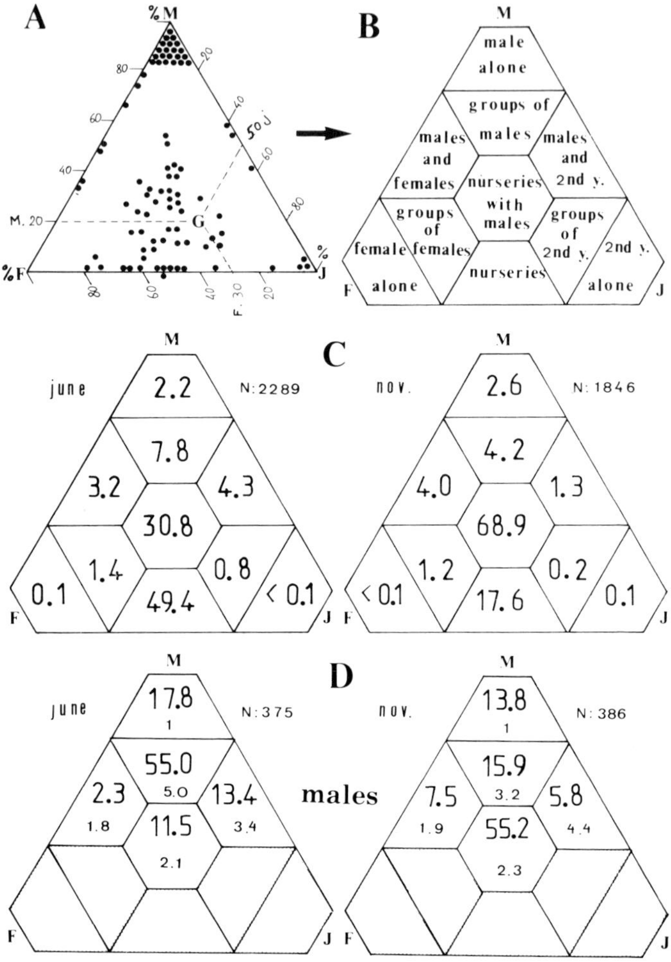

Figure 18.1: Age-sex Class Grouping Patterns of Chamois of the Ossau Mountain Area. A, Example of a triangular diagram showing the relative age-sex ratios of the groups (data were recorded during the 'flash-pointing' census of November 1980; see the text). B, Simple age-sex grouping diagram proposed for standard use. C, Relative distribution of all the chamois of the study area in the social group types set out in B. D. Distribution of the adult males (large numbers: percentage of individuals of the population involved in the corresponding group type, small numbers: mean size of such groups)

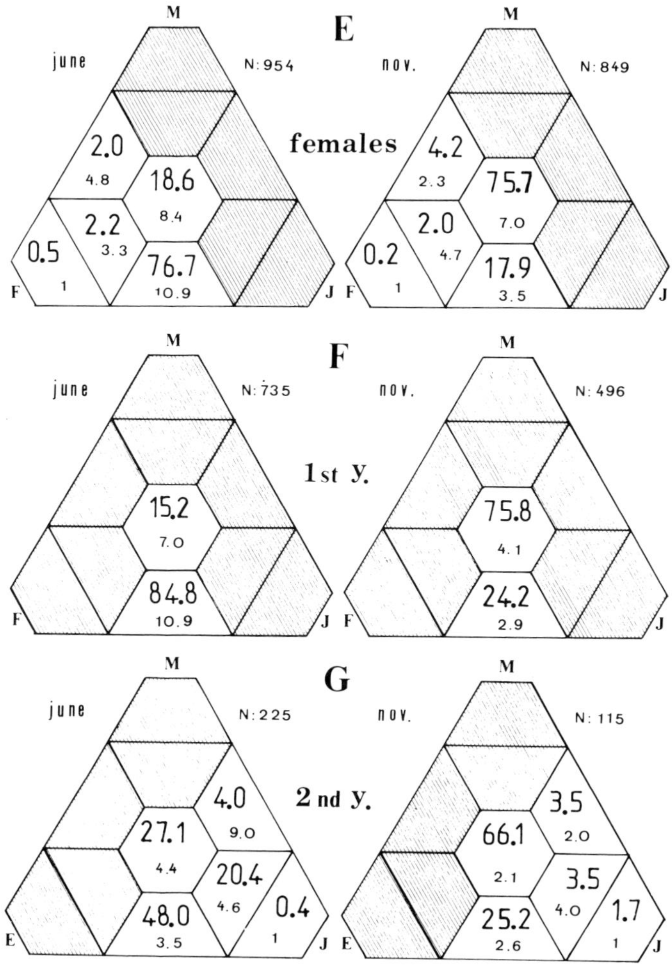

Figure 18.2: Age-sex Class Grouping Patterns of Chamois of the Ossau Mountain Area (Follows from Figure 18.1). E, Distribution of adult females. F, Distribution of 1st Y chamois. G, Distribution of 2nd Y chamois.

Social Links between the Different Age-sex Classes

In order to estimate the 'attraction' of one age-sex class (e.g. the adult females) for another one (e.g. the adult males), we calculated the 'Coefficient of Apparent Affinity' as:

$$\frac{\text{Number of males observed in groups with at least one female}}{\text{Total number of males observed}} \times 100$$

Table 18.2 sets out such values, based on the 'flash-pointing' data. The male-female bond increased from June to November; homogeneous class bonds (except female-to-female) decreased with the occurrence of mixed groups in the rutting season. Adult males and females grouped together very little in June (male-to-female coefficient: 14.0, female-to-male: 17.4). The most striking result is the strong male-to-male affinity in June (76.8).

Table 18.2: Coefficients of Apparent Social Affinity (see text) Between the Different Age-sex Categories of the Chamois of the Ossau Mountain Area

		Affinity of:			
		Males	Females	1st Y	2nd Y
for:					
Males	Jun.	76.8	17.4	13.8	41.3
	Nov.	59.0	76.9	76.8	69.1
Females	Jun.	14.0	92.7	100.0	62.8
	Nov.	62.4	95.2	100.0	90.0
1st Y	Jun.	10.0	91.7	95.1	44.8
	Nov.	55.3	92.2	91.1	86.4
2nd Y	Jun.	25.1	51.8	48.1	89.5
	Nov.	29.9	54.3	51.9	84.5

In order to study the tendency of adult males to group together, the 'periodic itinerary' method was used during one year on a small area supporting about 36 adult males (Table 18.3). Maximum cohesion was observed in June (mean group size: 11.9). In comparable conditions Schröder (1971) observed a maximum mean group size of 13.0 in July. The maximum adult male flock size observed in the Ossau area was 64 (old/adult males, except a few 3rd Y and 2nd Y).

Table 18.3: Year-round Evolution of the Group Sizes of About 36 Male Chamois Ranging in a Small Area

	Jan.–Mar.	Apr.–May	Jun.	Jul.	Aug.	Sep.	Oct.	Nov.–Dec.
Mean size	1.8	8.7	11.9	9.8	9.1	7.2	5.6	2.1
Max. size	4 (8)	34 (14)	32 (11)	31 (17)	25 (10)	20 (8)	24 (7)	7 (9)

In parentheses, the number of groups recorded.

Social Link Between the Different Individuals of the Same Age-sex Class

Through the 'fixed panoramic observation site' method, used during a period of 62 days in summer 1982, we recorded a females-kids group of about 45 chamois, in which 12 marked adult females were observed regularly. Figure 18.3 shows the links observed between these marked females during the morning grazing time in 22 consecutive days. Although some inter-individual bonds proved stable for a few days, the cohesion of the groups was continuously upset by a remarkable flow of individuals between groups. Nevertheless, this flow was restricted to a number of individuals ranging within a limited area. So, we define as a 'social unit' the number of individuals grouping together from time to time and forming recognisable but unstable groups.

Conclusion

At present, no comparison between different populations of *R.r. pyrenaica* is possible because of the lack of information from other areas of the Pyrenees.

Analysis of our results, however, shows three sociological features which seem typical of the Ossau massif dense population: very large nursery groups, large groups of adult males, and a high rate of mixing between social groups.

The observation that in a particular area the grouping pattern differs from that of other populations is not in itself an evidence

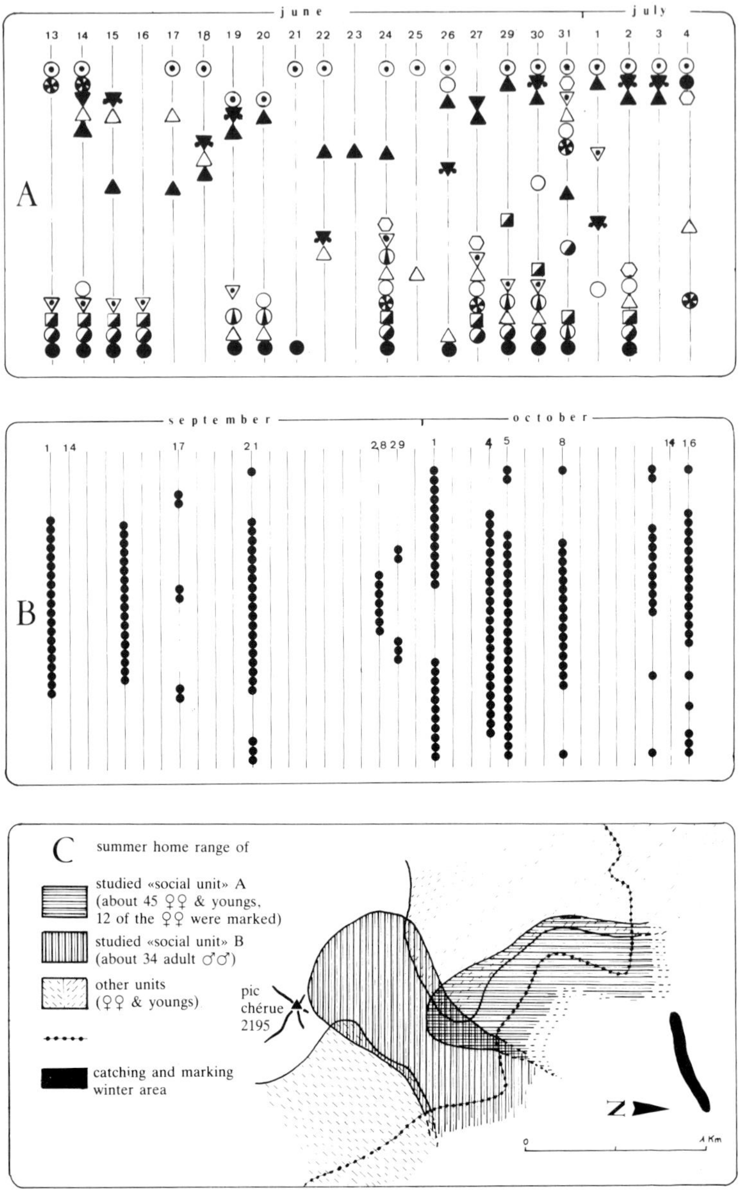

Figure 18.3: A, Evolution, during 22 Consecutive Days, of the Social Bonds Observed Between 12 Individually Marked Female Chamois in the Ossau Mountain Area. Each sign represents a particular female. For a given date, represented by a vertical line, each 'pack' of signs means that the

of over-population, but in future work we plan to compare our data with data from other Pyrenean populations in order to estimate to what extent the chamois social grouping patterns are density-dependent. If so, the social grouping patterns could provide the wildlife managers with an easily detectable sign of any imbalance between density of population and carrying capacity of the habitat.

References

Altmann, J. (1974) 'Observational Study of Behaviour: Sampling Methods', *Behaviour*, *49*, 227-67

Berducou, C., Besson, J.P. & Gardes-Moniteurs du PNPO (1982) 'Dynamique des Populations d'Isards du Parc National des Pyrénées Occidentales, de 1968 à 1981', *Acta Biolog. Montana*, *1*, 153-75

Krämer, A. (1969) 'Soziale Organisation und Sozialverhalten einer Gems-population der Alpen', *Z. fur Tierpsychologie*, *26*, 889-964

Lovari, S. (in press) 'Herding Strategies of Male Abruzzo Chamois on the Rut', *Acta Zool. Fenn.*

Schröder, W. (1971) 'Untersuchungen zur Ökologie des Gamswildes in einem Vorkommen der Alpen', *Z. f. Jagdwiss.*, *17*, 114-68; 197-235

corresponding individuals were observed in the same group (including also unmarked individuals, not represented here). B, Aggregation patterns of the males of a 'social unit' (see text). C, Summer ranges of the 'social units' mentioned in A and B, and of two other units

Plate Section

Plate 1 A prime male of Tatra chamois (*Rupicapra rupicapra tatrica*) in its winter coat. (*M. Blahout*)

Plate 2 A female Tatra chamois nursing her kid. The stocky complexion of this subspecies may be an adaptation to life in a cold climate. (*M. Blahout*)

Plate 3 Behavioural, morphological, anatomical and electrophoretical studies have recently demonstrated that the Abruzzo, or Apennine, chamois can be grouped as a separate species (*Rupicapra pyrenaica ornata*) together with the chamois of the Pyrenees (*R. p. pyrenaica*) and the Cantabrics (*R. p. parva*), separate from all other chamois (*Rupicapra rupicapra*). (*S. Lovari*)

Plate 4 The Asiatic serow (*Nemorhaedus sumatraensis*) is an endangered goat-antelope living in forested, rocky areas. Its biology is still totally unknown. (*S. Lovari*)

Plate 5 Gorals (*Nemorhaedus* spp.) are small Asiatic mountain ungulates living singly or in bands of up to 12 individuals. Photographs of this secretive species in the wild are extremely rare. (*C. Ngampongsai*)

Plate 6 The numbers of Spanish ibex (*Capra pyrenaica*) declined alarmingly during the nineteenth and early twentieth centuries. Only in 1950 was a conservation programme started. (*C.L. Alados*)

Plate 7 A herd of moufflon rams (*Ovis musimon*) in March. Note the white saddle shown by mature rams. (*C. Novoa*).

Plate 8 The Alpine ibex (*Capra ibex*) was on the verge of extinction early in this century, surviving only in the Gran Paradiso mountains, Italy. Effective management measures have now saved this species by re-introducing it in many other Alpine areas. (*F. Francisci*)

Plate 9 The lush Alpine meadows on the south-facing slopes of the Vanoise massif, French Alps, are important summer grazing areas for the ibex. (*G. Wiersema*)

Plate 10 Ibex winter range in the Gran Paradiso National Park, Italy. (*G.J. den Daas*)

Plate 11 Do ibex occur in trees? The surprising answer is: Yes! (*G. Wiersema*)

Plate 12 Two male Alpine ibex sparring. On the rut, in December, the Alpine valleys resound with the clashing noise of the fighting males. (*Arch. Abt. Wildf. Oberammergau*)

Plate 13 Male Sardinian moufflons in their sleek, short summer coat. This species has been introduced for hunting purposes into several areas where it is now adversely affecting other local ungulate species. (*D. Ruiu*).

Plate 14 Outside the rut, male and female Alpine ibex tend to occur in unisexual bands, or even in separate areas. A male group (in the picture). (*B. Georgii*)

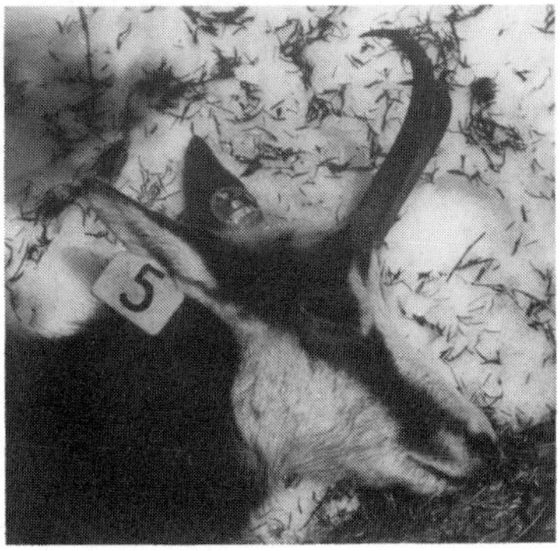

Plate 15 A narcotised 11-year-old Alpine chamois, showing a sun-powered radio-transmitter (on left ear) and a plastic tag (on right ear) inserted for scientific purposes (*J. Hamr*).

Plate 16 Landsat satellite images have been used in ibex habitat studies in the Alps. This image of 23 May, 1976, shows the spring snowcover situation in the Vanoise (V) and Gran Paradiso (GP) massif on the French-Italian border. Computer processing assists in finding seasonal habitats, e.g. winter range on snowfree, vegetated slopes. The sequential onset of vegetation growth can also be detected by processing a green biomass index for the different seasons. (*NASA/ESA. Project Ibex*)

Plate 17 Lambs of Alpine ibex are agile, attractive creatures. Twins are very rare. (*B. Georgii*)

Plate 18 A yearling male Alpine chamois, still drowsy upon being narcotised, with a radio-collar. (*J. Hamr*)

Plate 19 In winter the chamois tend to scatter in steep, forested slopes. Groups such as this (*photo*) are relatively rare. (*S. Lovari*)

Plate 20 A male Alpine ibex horning vegetation. Little is known of the behaviour of ibex, as studies on this species have been mainly concerned with ecological aspects. (*Arch. Abt. Wildf. Oberammergau*)

Plate 21 The infectious keratoconjunctivitis develops through four stages. The animal may recover from the first two, but it is bound to suffer blindness in the third and fourth ones. (*Gruppo Ibex – Turin*)

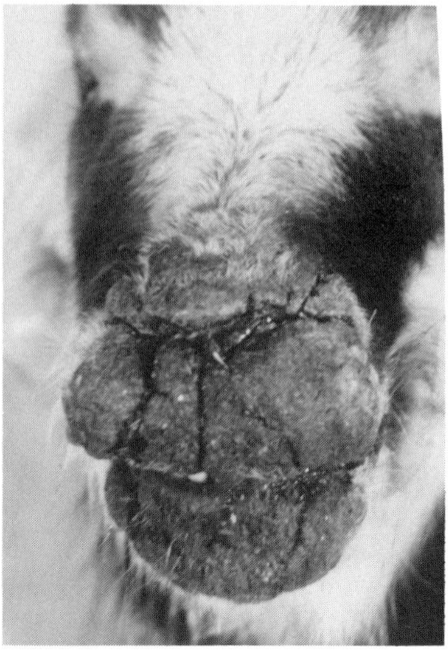

Plate 22 The contagious ecthyma is a disease affecting the mucosa tissues of chamois, to which it causes painful wounds. (*Gruppo Ibex – Turin*)

Plate 23 The mange, or scabies, is probably the most important disease of chamois. It is caused by a mite, *Sarcoptes rupicaprae*. (*Arch. Abt. Wildf. Oberammergau*)

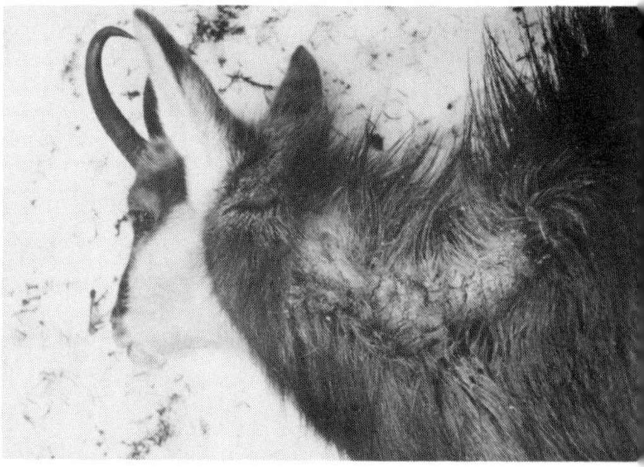

Plate 24 A 3-year-old male Abruzzo chamois. The only remnant population of these chamois (350-400 individuals in all) occurs in some mountains of the Abruzzo National Park, Italy, and it is exceptionally healthy. Its reintroduction is being planned for several other Apennine areas. (*S. Lovari*)

PART THREE

Management and Conservation

CHAPTER 19

Management of Mountain Ungulates

W. SCHRÖDER
Institute of Wildlife Research, University of Munich, Amalienstrasse 52, D-8000 München 40 (FRG)

Abstract

The various management problems are reviewed considering the species of the subfamily Caprinae. Six major management objectives are discussed: 1. conservation, 2. harvesting, 3. prevention of damage, 4. disease control, 5. extension of the natural range, and 6. natural regulation in parks. The effective protection of species and populations is seen to be the most important management task on a worldwide scale. Wildlife managers must receive better training in the political and administrative aspects of implementing mountain ungulate conservation within the framework of general land use.

Introduction

When I was asked to prepare a paper on this topic, I searched for some particular aspect that is unique to managing mountain ungulates. There is none I could see. All the problems are familiar in ungulate management or even wildlife management in general, the differences being only matters of degree. One factor that often applies, but it is not unique to mountain ungulates, is the life in extreme habitats. Aridity, short growing seasons, deep snows or rugged terrain and fragmentary habitat, all have ecological consequences for population dynamics. The job of managing mountain ungulates, as I understand it, comprises the definition of relevant goals and also the process of steering the species concerned, its habitat and the people involved towards the defined objective. Since the relevant literature is very extensive, no attempt is made to review all of it. The problems have been grouped into 6 different categories according to management objectives. These are: 1. conservation, 2. harvesting, 3. prevention of damage, 4. disease control, 5. extension of the natural range, and 6. natural regulation.

Mountain ungulates are not a particular taxonomic group. 'Born of the Pleistocene, at home among pulsating glaciers and windflayed rocks, the animals have survived and thrived, the harshness of the environment breeding strength and resilience which the lowland animals often lack' (Schaller, 1977) Practically all of them belong to the subfamily of Caprinae, which includes three tribes: Rupicaprini, Ovibovini, Caprini (Table 19.1). All of these animals, except the musk-ox, can climb well and use rocks or rugged ground for escape terrain.

There are other ungulates at high altitudes, like the vicuña in the Andes, or the red deer, summering up to 3,000 m in the Swiss Alps, well above the timberline, and I have seen caribou high up in the mountains of Jasper National Park in Canada. Since these species are not particularly adapted for climbing, or since they occupy winter ranges off the mountains, they are excluded, like the musk-ox, from further consideration.

Conservation

Like so many other creatures, mountain ungulates have suffered with the ascent of modern man. There is not a single species among them that was not reduced markedly at some point after having come in close contact with man. Some, like the chamois, have recovered well, others, like the bighorn sheep and the alpine ibex, partially, but many are worse off than ever before.

Protection of Viable Populations

Most mountain ungulates in Asia are many fewer than the mountains could support. The species are too numerous to list. George Schaller has described their bleak situation in his two revealing books (Schaller, 1977, 1980). The reasons are manifold and interrelated. Many arid mountains have lost the former vegetation because of overgrazing by goats and sheep, and

Table 19.1: Subfamily Caprinae

Tribe	Genus	Species	Common name
Rupicaprini	*Nemorhaedus*	*goral*	Goral
	Capricornis	*sumatraensis*	Asiatic or mainland serow
		crispus	Japanese serow
	Oreamnos	*americanus*	Rocky Mountain goat
	Rupicapra	*rupicapra*	Chamois
Ovibovini	*Ovibos*	*moschatus*	Musk-ox
	Budorcas	*taxicolor*	Takin
Caprini	*Ammotragus*	*lervia*	Aoudad or Barbary sheep
	Pseudois	*nayaur*	Bharal or blue sheep
	Hemitragus	*jemlahicus*	Himalayan tahr
		hylocrius	Nilgiri tahr
		jayakeri	Arabian tahr
	Capra	*aegagrus*	Wild goat
		ibex	Ibex
		falconeri	Markhor
		pyrenaica	Spanish ibex
		cylindricornis	East Caucasian tur
		hircus	Domestic goat
	Ovis	*canadensis*	Bighorn sheep
		dalli	Thinhorn sheep
		nivicola	Snow sheep
		ammon	Argali
		orientalis	Urial
		aries	Domestic sheep

protective forests were turned into timber and burned as firewood.

Competition from domestic livestock is a major problem. Where livestock is, people are not far away, and they kill many mountain ungulates, for various reasons, legal or illegal. With the breakdown of the traditional (aristocratic) authorities, their well-protected hunting preserves are gone. Some of the recent endeavours by international organisations, like the FAO, to establish some parks are often buried in political turmoil, as in Afghanistan right now (C.C. Shank, personal communication). There is, however, an urgent need to create parks protecting the nuclei of possible future populations. In the long run, though, parks alone are not sufficient to preserve these large species, if our ideas about the insularisation are correct (Soulé, Wilcox &

Holtbyx, 1979). It is important to develop land use patterns integrating wildlife management into range and forest management — no easy task in the face of increasing human populations, old grazing traditions and vast empty mountains, where the shot of a rifle echoes unheard. Schaller (1980) writes: 'But the great age of mammals in the Himalaya need not be over unless we permit it to be. For epochs to come the peaks will still pierce the lonely visitors, but when the last snow leopard has stalked among the crags and the last markhor has stood on a promontory, his ruff waving in the breeze, a spark of life will have gone, turning the mountains into stones of silence'.

What is true for the ungulates in the Himalaya and the Hindukush can also be said for the Walia ibex in the Simien Mountains of Ethiopia. Here also various human activities encroach on a small national park, in this case with about 300 ibex (Nievergelt, 1981).

A threat of a different nature is the exploration for coal and other natural resources in remote mountains of Canada (Pendergast & Bindernagel, 1977). Mountain goats are likely to decline in areas of easy access because of concentrated hunting pressure (Figure 19.1). But in a country with high standards of wildlife management, such problems can be anticipated and solved quite quickly.

Protection of Subspecies

A threat to certain subspecies exists, not from extinction but from mixing with introduced populations of other origin. This is the case for chamois, for example, where Lovari (1984, and personal communication) has gathered information (Figure 19.2). In Yugoslavia the subspecies *R.r. balcanica* has meanwhile intermixed with several populations of *R.r. rupicapra* of alpine origin. In Czechoslovakia the alpine subspecies was taken to found new populations very close to the range of *R.r. tatrica*. In France also, alpine chamois were introduced near the Chartreuse massif, where *R.r. cartusiana* is dwindling to dangerously low numbers. Actions like these will lead to the loss of genetic identity of defined and isolated subspecies.

Preventing Loss of Genetic Variation

Genetic variation, the evolutionary potential of populations, is in jeopardy when populations stay at low numbers for long periods.

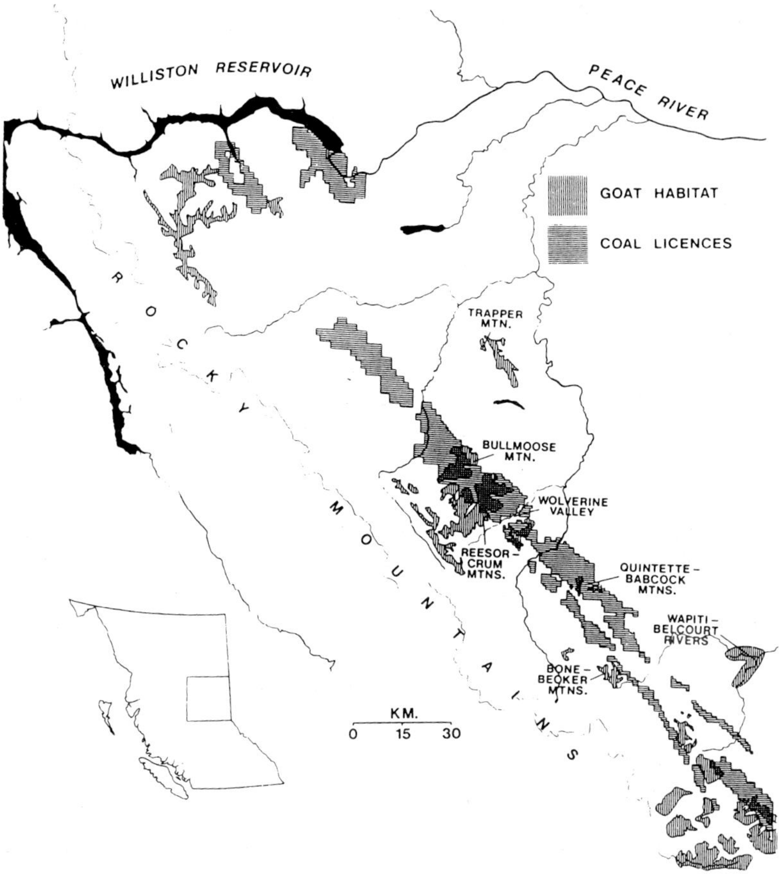

Figure 19.1: Goat Habitat and Coal Licences in British Columbia. Mountain goat numbers declined drastically in areas accessible by road. Habitat destruction does not seem to be as important a factor in losses as is easy human access (*from* Pendergast & Bindernagel, 1977)

Many populations are in such a state; for genetic reasons population size should be brought quickly to a level of several hundred animals. Otherwise such bottlenecks result in large losses of genetic variation (Frankel & Soulé, 1981). Introduced populations of ibex or mountain goat in marginal habitat that persist at low numbers are therefore not desirable.

Genetic principles must also be considered when establishing new populations with few animals, minimising an undesirable

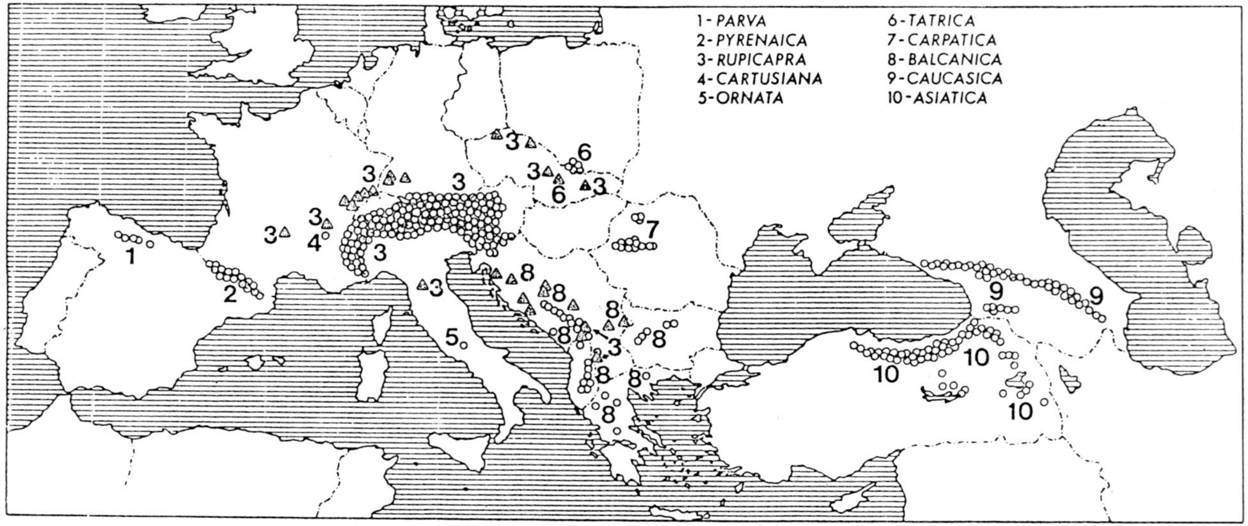

Figure 19.2: Natural Distribution (circles) and Introduced Populations of Chamois (triangles). Most populations were founded with the Alpine subspecies (*R.r. rupicapra*) even when within the range of other subspecies (*from* Lovari, 1984).

founder effect. In the case of alpine ibex the founder effect is particularly severe: almost all populations go back to ancestors that were bred in a Swiss pen for some time. Because animals were taken from newly established populations to found further ones which became the origin of further populations, there is a multiple founder effect. Genetic research and the application of genetic principles to conservation have made such progress over recent years that ignorance will be no excuse in the future (Schonewald-Cox *et al.*, 1983).

Re-establishing Lost Populations

Some mountain ungulates are such poor colonisers that many introductions were necessary to re-establish lost populations, the bighorn sheep and alpine ibex being the most well-known examples. The alpine ibex was brought back from a single small population of several dozen animals in the last century to around 25,000 individuals and about half the former range (Figure 19.3) (v. Elsner-Schack, 1982).

In contrast to chamois, ibex can only survive in mountains with little rain and little snow. So populations will always colonise certain mountains and not others. Recently a method has been developed including Landsat-imagery to find suitable ibex habitats (Wiersema, 1983). The problem of finding habitats for reintroduction becomes more pronounced, where the best and obvious habitats have already been reclaimed by ibex. The method can also be used for find proper range when establishing reserves in other regions for other mountain ungulates.

Harvesting

Methods of hunting for meat or trophies are as varied as the mountains. Tradition is still the most important determinant even in countries where wildlife management is practised. In the Alps this has led to a mosaic of management approaches, heterogeneous even within Switzerland or within Austria.

Some Principles of Population Harvesting in Mountainous Habitats

Harvesting regimes have evolved out of long experience. Many of them are practicable and well suited to the needs of the country. They are not usually based on ecological theory and are therefore not adaptable to changing situations. There is, for example, a wealth of literature on how to manipulate the age distribution of populations through harvesting in order to maximise adult bucks in chamois. There is, however, not enough consideration of the population dynamic consequences of such practices because this is much more difficult to project. The CIC (Conseil Internationale de la Chasse), during a meeting on chamois, recently came up with the general recommendation to shoot more animals in young age classes. Such a statement may not be supported by great insight into processes of population ecology.

We have recently analysed chamois population reaction to various harvesting regimes under different climatic conditions through computer simulation. The model used was developed in a study about the population regulation system of chamois mostly in the Berchtesgaden National Park, Bavaria (Schröder *et al.*, 1983). Some of the more general insights can be summarised as follows. Not all are new to population managers:

a. There is an increasing stochastic and density-independent effect with increasing average amount of snow in the various mountains (Figure 19.4).
b. Constant sustained yield harvests of different magnitude lead to different equilibrium populations (Figure 19.5b and c)
c. High cropping reduces environmentally induced stochastic influences on population variation (Figure 19.5c).
d. In mountain ranges with different amounts of snow, constant sustained yield harvests of the same magnitude result in different equilibrium populations (Figure 19.6).
e. Within limits, the harvesting of young animals instead of adults does not have the same reducing effect on the population (Figure 19.7).

Looking at present-day harvesting practices not all of these statements are obvious or trivial. There is much room for improvement. For mountain ungulates in general, density-independent effects have been considered more often than in other ungulate species. This is why harvest rates in some cases may be low as in chamois, ibex or mountain goats. Hunting losses and natural mortality may not be acting in a compensatory fashion but more or less adding up (Kuck, 1977). This is why many mountain ungulates cannot be harvested as heavily as deer.

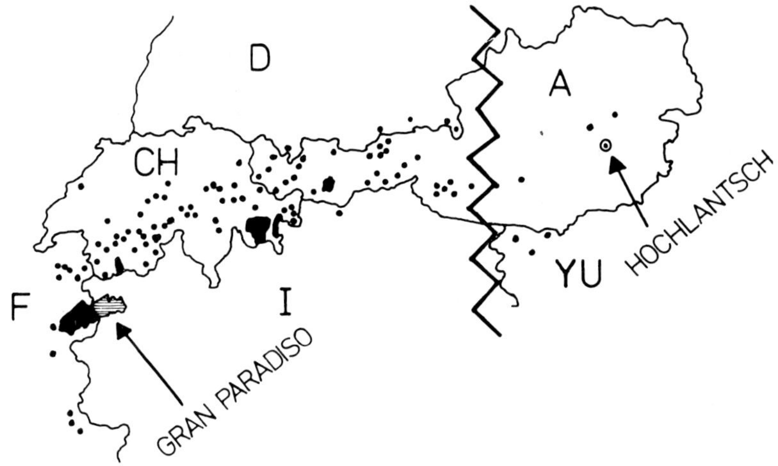

Figure 19.3: Present Distribution of Ibex in the Alps. All populations go back to the Gran Paradiso National Park. There are no records of post-Pleistocene ibex in the Eastern Alps to the right of the zig-zag line

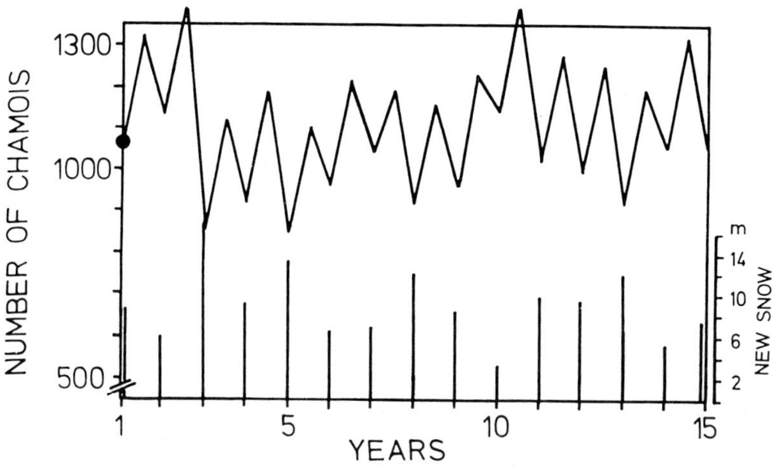

Figure 19.4: Simulation of an Unharvested Population of Chamois in the Berchtesgaden National Park. The new snow depth is great and highly variable on the north slope of the Alps. A summer and winter population is shown for each year

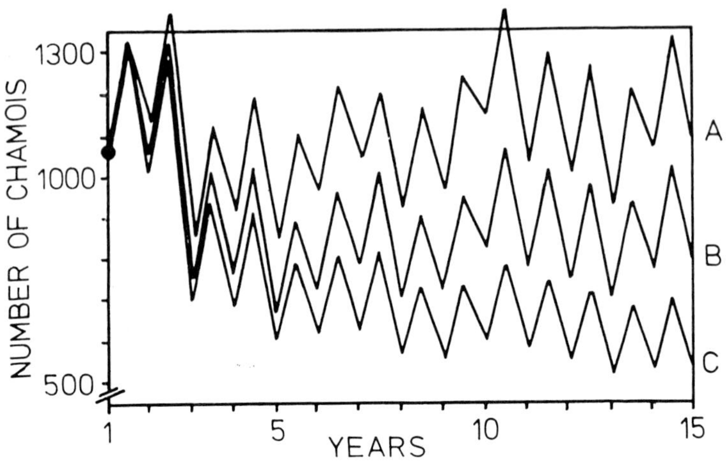

Figure 19.5: Simulating the Effect of Sustained Yield Harvesting. A, unharvested population as in Figure 19.4; B, population with constant harvest of 50 animals of both sexes; C, population with a constant harvest of 100 animals of both sexes

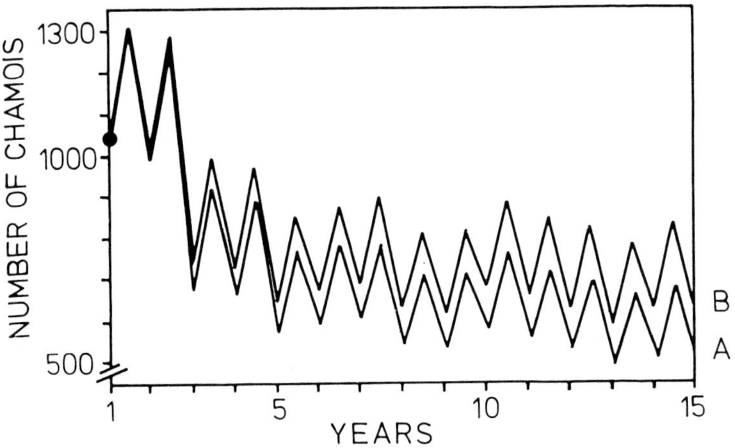

Figure 19.6: Simulating the Effect of Different Winter Conditions and Harvest. A, population subject to the snow of Figure 19.4 subjected to a constant harvest of 100 animals of both sexes; B, same population and same harvest, but average amount of snow is less

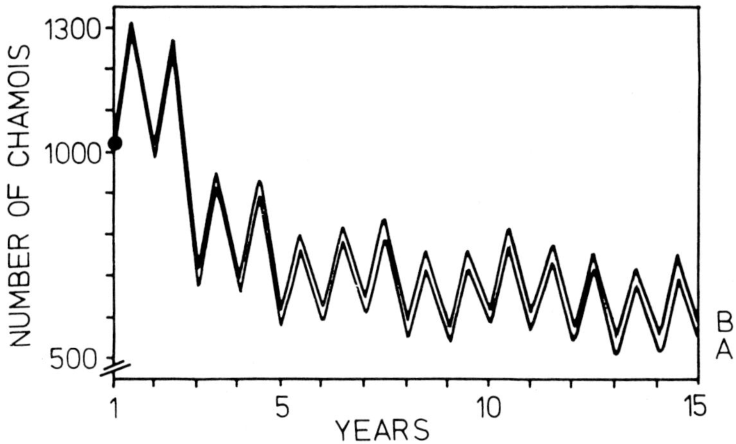

Figure 19.7: Simulating the Effect of Shooting Young Chamois Instead of Adults. A, population subjected to a constant harvest of 100 adult animals of both sexes; B, population when 90 adult and 10 young of the year of both sexes are harvested

Eruptive Populations

Populations introduced into new habitats may increase, overshoot and demonstrate different kinds of cycles, depending on population parameters. This can cause temporary phases of overpopulation with undesirable effects on the environment. This can be prevented when cropping, set in at the right time, prevents the overshoot, leading the population to equilibrium in a less spectacular way.

An unusually productive population of alpine ibex, introduced outside the former range, grew rapidly at 16 per cent per annum, competing effectively with chamois and putting much pressure on the vegetation (Kofler, 1983). The population is now cropped regularly and we will calculate the harvest using computer simulations to avoid overshooting (Kofler & Schröder, 1985).

Manipulating Behaviour

Bighorn sheep, ibex and probably other species as well are very traditional in their movements within their home ranges (Geist, 1977). Shooting animals not only has a numerical effect on the population but also influences distribution and migrations. This is particularly true where females are harvested. Behaviour will have to be taken into consideration where harvesting ibex becomes an important population regulation factor.

Habituation of animals to road-traffic and people has sometimes been observed in North American National Parks. Where relations are too intimate — sheep being fed candy bars out of car windows — this habit should be broken on both sides. For a European wildlife biologist such behaviour on the part of the sheep is rather surprising since habituation cannot be observed to this degree in European species. This may have to do with the negative experience that our species have had with man for a much longer period. This difference can, by the way, also be observed in wolves and bears.

Control of Damage

Luckily this is not a great problem with mountain ungulates. Chamois and ibex sometimes hurt the young trees, where special effort is made to push the timberline back up to where it was, before pasturing in the Middle Ages lowered it. Chamois, the most numerous and adaptive of all, have colonised new terrain, extending their former range far into the forested foothills on the north slope of the Alps. Wildlife browsing prevents the regeneration of the spruce-fir-beech forest in many such areas, chamois being one of the species involved. To solve this problem a proposal was presented for delineation of chamois habitat according to rock distribution. Chamois should be removed from range without rocks. Increasing the harvest in the main ranges will also lead to withdrawal of chamois from the periphery or marginal habitat, because it is assumed that they fill the habitat beginning at the higher quality ends (Schröder & v. Elsner-Schack, 1982).

Control of Disease

Some of the species are suffering great losses from diseases. The pneumonia complex in bighorn sheep and scabies in chamois are among the most important with regard to population consequences. Much research has been done on both phenomena (e.g. Post, 1971, 1976; Onderscheka, 1982) but there are still open questions. Bighorn sheep currently suffer heavy losses in spite of overall low population numbers, when compared to the pristine distribution. Locally, high population densities in relation to food resources can apparently lead to density dependent outbreaks. Harvest of females to reduce the population has been recommended and practised successfully on Pike's Peak in Colorado (Morris, 1976).

Not all chamois populations seem to harbour the sarcoptic mite, the parasite causing the disease. Scabies has spread to new mountain ranges in Austria, Italy, Bavaria, Germany and Slovenia (Yugoslavia) in recent years (Figure 19.9). There is a correlation of the probability of outbreak and population increase of chamois as indicated through steadily climbing harvest statistics. All of the countries sharing the Alps have seen an increase of chamois over the last 30 years following changes in management (Figure 19.8). This has preconditioned the populations to a variety of diseases. With scabies, infected populations remain at a much lower equilibrium because of the controlling action of the parasite (see also Anderson, 1979). No effective cure has yet been found.

Extending the Former Range

Some species like the mountain goat, the ibex and chamois were introduced into mountains close to their former range, but where they did not occur naturally. This can be seen as an artificial extension of the natural range. Some transplants are of different nature, because of the transport to different zoogeographic areas, where they are called exotics. The barbary sheep in New Mexico or the chamois in New Zealand are examples.

At present the Rumanian government has acquired some Alpine ibex from the Gran Paradiso National Park to populate the Carpathians. This mountain range is not good ibex habitat in my opinion, because of heavy rain and snow; besides, there were no ibex there even in the Pleistocene epoch.

Some years ago the Bavarian Prime Minister was about to give some chamois to the state of Colorado to be released in the mountains there upon the request of military personnel who had

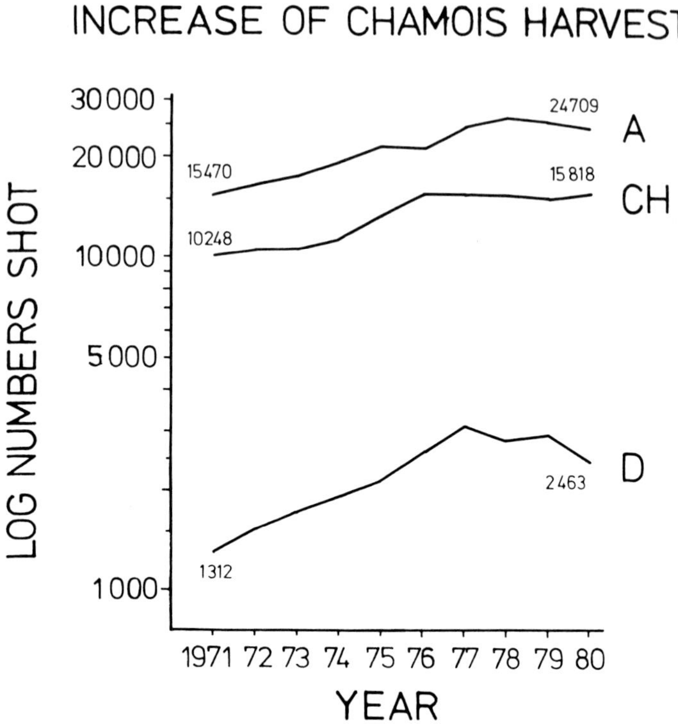

Figure 19.8: Increase of Chamois Harvest in Three Different Countries in the Alps. A: Austria, CH: Switzerland; D: Federal Republic of Germany

become familiar with chamois and chamois hunting in the Alps. With the authoritative help of V. Geist from Canada I was able to stop this ill-advised project.

Some of the species introductions were made in improper habitats, such as the ibex in mountain ranges with too much snow or into forested mountains of low altitude. The same is true for chamois in the Black Forest, which has no open range worth mentioning. Some Black Forest chamois were given to the French people to establish a population across the Rhine in the Vosges Mountains, which are also completely forested. The population has in the meantime increased to more than 1,000 animals, doing much damage to the forest. Because, under the French occupation, large tracts of the Black Forest were cut and transported across the Rhine some people thought of this as an act of late revenge.

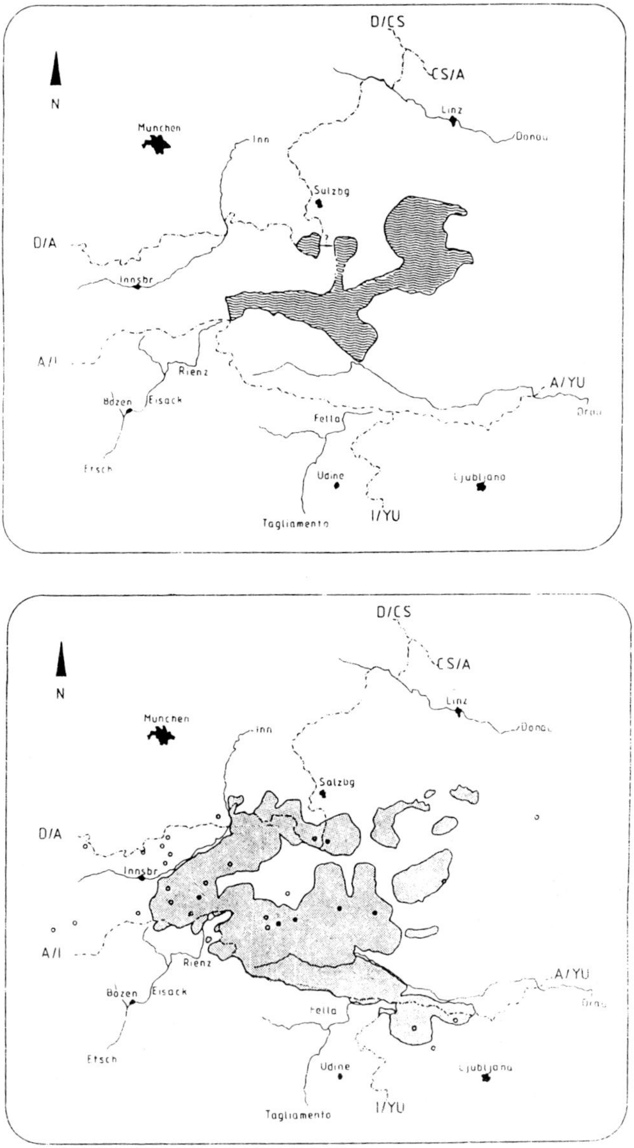

Figure 19.9: Scabies in Chamois around 1917 (left) and in 1980 (right). Scabies is still spreading to new populations (*from* Miller, 1982)

Maintaining Natural Population Regulation in Parks

The maintenance of natural processes, including natural population regulation, is often a management objective in National Parks. Nothing seems unusual to people when this has been practised for long periods. In Europe the Swiss National Park does not control ibex or chamois populations and the Abruzzo Park does not control its chamois. There is a controversial discussion in other parks, such as the Berchtesgaden National Park in Bavaria, whether this can be done without detriment to the park and mountains surrounding it. It is, as Caughley (1981) has shown for Africa, not an ecological problem, although many of the arguments used are ecological and therefore beside the point. In some countries, allowing natural regulation, even in parks, requires, as Caughley phrased it, a complete mental reorientation, the crossing of a mountain range of mind.

Conclusions

Looking at the overall picture, effective protection of species and populations is the most important task in management of mountain ungulates. We have to train wildlife managers not so much in ecological theory but to teach them to clear a path through the political and administrational jungle if we are effectively to implement mountain ungulate conservation into land use practices. Otherwise there will be more 'stones of silence'.

Some of the luckier countries and wildlife managers can afford to spend time on difficult questions of population management and harvesting. Scientists and administrators must learn to cooperate more closely even when they have developed traditionally along different lines. There is much less institutionalised cross-fertilisation in countries of the Old World compared to the New World with regard to the various state and governmental wildlife services.

We have to define standards for the maintenance of subspecies and the extension of the range of species through introductions. In the minds of many people with a sense of naturalness, there is a limit to what should be done. This again is not an ecological question. The mountain goat conveys the wildness of the Rocky

Mountains and so does the chamois of some Old World mountain ranges. This image is precious and important to some people. It must not be destroyed.

References

Anderson, R.M. (1979) 'The Influence of Parasitic Infection on the Dynamics of Host Population Growth' in R.M. Anderson & T.A. Taylor (eds.) *Animal Population Dynamics*, Blackwell, Oxford, pp. 245-80

Caughley, G. (1981) 'Overpopulation' in P.A. Jewell & S. Holt (eds.) *Problems in Management of Locally Abundant Wild Animals*, Academic Press, New York, pp. 7-19

Elsner-Schack, I. von (1982) 'Zur Wiedereinburgerung des Steinbocks in den gesamten Alpen' in H. Kofler (ed.) *Der Steinbock*, Wildbiologische Gesellschaft Munchen, pp. 9-20

Frankel, O.H. & Soulé, M.E. (1981) *Conservation and Evolution*, Cambridge University Press, Cambridge

Geist, V. (1977) 'On the Management of Mountain Sheep: Theoretical Considerations' in J.B. Trefethen (ed.) *The Wild Sheep in Modern North America*, Winchester, New York, pp. 77-105

Knaus, W. & Schröder, W. (1983) *Das Gamswild*, Paul Parey Verlag, Hamburg and Berlin

Kofler, H. (ed.) (1983) *Der Steinbock: Ansprüche, Einbürgerung, Bejagung,* Tagungsbericht Wildbiologische Gesellschaft, Munchen

Kofler, H. & Schröder, W. (in press) 'Coexistence and Competitive Exclusion of Ibex and Chamois', *Acta Zool. Fenn.*

Kofler, H. & Schroder, W. (1985) 'Harvesting an Atypical Ibex Population: A Management Plan' in S. Lovari (ed.) *The Biology and Management of Mountain Ungulates*, Croom Helm, London, pp. 212-15

Kuck, L. (1977) 'The Impacts of Hunting on Idaho's Pahsimeroi Mountain Goat Herd' in W. Samuel & W.G. Macgregor (eds.) *Proc. 1st Int. Mountain Goat Symp.*, Kalispell, Montana, pp. 114-25

Lovari, S. (1984) *Il Popolo delle Rocce*, Rizzoli, Milan

Miller, C. (1982) 'Ausbreitung der Gamsräude in Bayern und Dynamik der befallen Bestände', Dipl. Arbeit, Universität Munchen

Morris, J. (1976) 'Justifications for Pike's Peak either Sex Sheep Season' in E. Decker (ed.) *Trans. 2nd North American Wild Sheep Conf.*, Denver, Colorado, Dept. of Fishery & Wildlife Biology, Fort Collins, pp. 82-5

Nievergelt, B. (1981) *Ibexes in an African Environment*, Springer Verlag, Berlin, Heidelberg, New York

Onderscheka, K. (1982) 'Aktueller Stand der Gamsraudeforschung', *Vortrag Intern. Jagdkonf.* 22-23 May 1982, Matrei/Osttirol, pp. 36 (Mimeographed)

Pendergast, B. & Bindernagel, J. (1977) 'The Impact of Exploration for Coal on Mountain Goats in Northwestern British Columbia' in W. Samuel & W.G. Macgregor (eds.) *Proc. 1st Int. Mountain Goat Symp.*, Kalispell, Montana, pp. 64-8

Post, G. (1971) 'The Pneumonia Complex in Bighorn Sheep' in E. Decker (ed.) *Trans. 1st North American Wild Sheep Conf.*, Colorado State University, Dept. of Fishery and Wildl. Biol., Fort Collins, pp. 98-106

—— (1976) 'Diagnostics and Diseases of Wild Sheep' in E. Decker (ed.) *Trans.*

2nd North American Wild Sheep Conf., Denver, Colorado, Dept. of Fishery & Wildl. Biol., Fort Collins, pp. 61-5

Schaller, G.B. (1977) *Mountain Monarchs*, The University of Chicago Press, Chicago & London

—— (1980) *Stones of Silence*, Viking, New York

Schonewald-Cox, Ch., Chambers, S.M., Macbryde, B. & Thomas, L. (1983) *Genetics and Conservation*, Addison-Wesley, Reading

Schröder, W. (1982) 'Zur Jagdlichen Planung in Steinbockpopulationen' in H. Kofler (ed.) *Der Steinbock*, Wildbiologische Gesellschaft Munchen, pp. 37-49

—— & Elsner-Schack, I. von (1982) 'Gams und Gebirgswald', *Fortstwiss. Centralb.*, *101*, 80-92

——, Elsner-Schack, I. von & Schröder, J. (1983) 'Die Gemse', *Jahrbuch des Vereins zum Schutz der Bergwelt*, *48*, 33-70

Soulé, M.E., Wilcox, B.A. & Holtbyx, C. (1979) 'Benign Neglect: A Model of Faunal Collapse in the Game Reserves of East Africa', *Biol. Conserv.*, *15*, 259-72

Wiersema, G. (1983) *Ibex Habitat Inventory and Mapping in the European Alps*, (Mimeographed) pp. 1-8

CHAPTER 20

Management and Conservation of the Sardinian Moufflon (*Ovis musimon* Schreber). An Outline

F. CASSOLA
Vice-President, WWF-Italy, Via F. Tomassucci 12, 00144 Roma (Italy)

Abstract

The present status of the Sardinian moufflon (*Ovis musimon* Schreber 1782) in its original range (central-eastern Sardinia) is briefly reviewed. In comparison with the 1960s, numbers appear to have recovered quite well, reaching over a thousand animals, but poaching, road building and increasing tourism still represent a major threat for moufflon survival. The importance of establishing the long-awaited Gennargentu National Park for the conservation of the moufflon environment is consequently emphasised.

Re-introductions into other suitable areas of Sardinia are also recommended. Introductions, on the contrary, appear to be acceptable only in particular situations, as a means of enriching a depleted environment in order to improve the conservation of the area. Finally, introductions ought always to be firmly avoided when other ungulates (e.g. the chamois), which may be adversely affected by the presence of the moufflon, occur in the releasing area.

Introduction

The Sardinian moufflon (*Ovis musimon* Schreber 1782) can be regarded as one of the most interesting species of Mediterranean wildlife. Its phylogeny, systematics, distribution and conservational importance have been discussed in a previous report (Cassola, 1976). It may be recalled here that the species, once widely spread over the whole of Sardinia, has showed a severe decline in recent decades, and has disappeared from most previously inhabited areas. Its present range is restricted to a few separate areas of central-eastern Sardinia, especially in remote mountainous regions of the Gennargentu and Supramonte massifs. In addition, two introduced populations occur at Capo Figari near Olbia (approximately 200 animals) and on the Asinara Island, north-western corner of Sardinia (approximately 300 animals).

The Corsican population has also severely declined. Fifteen years ago no more than 100-200 animals were left, divided in two main herds in the Bavella and Asco Reserves (Pfeffer, 1967). In recent years numbers appear to have recovered quite well, especially after the creation of the Corsican Regional Park (Leenhardt, personal communication, 1982; Dubray, in press), but the animals are still divided into three geographic isolates (Asco, Lonca, Bavella) whose preservation is not yet secured (Demeautis, in press). The Corsican moufflons differ from those of Sardinia, since most of the ewes are horned, whereas most of the Sardinian ones are not.

Besides these two Mediterranean islands, moufflons have been introduced in many other countries, mainly for hunting or ornamental purposes. They have proved to be very robust and adaptable animals, by establishing themselves successfully in many different habitats. Nowadays there are many thousand moufflons in most European countries, and even in places remote from their natural range such as New Zealand, Hawaii and the Kerguelen Islands.

Numbers

The status of the Sardinian moufflon in its original range has recently caused concern. The first tentative census, based mainly on secondhand information, was made by Reinhardt & Schenk (1969), who reported very small numbers, i.e. 300-360 individuals in the whole original range (50-80 at Mt. Albo, 150-180 in the Supramonte Massif, and 100 in the Gennargentu/Mt. Tonneri area). Similar low numbers were also given by Schenk (1972, 1975), Massoli Novelli (1973), Van Haaften (1973), Silvestri (1974) and Perco (1981), nevertheless these are probably underestimates, since in more recent reports (Schenk 1976, 1980) numbers appear to be much higher.

In fact, recent information (Ruiu & Congiu, *in litt.*, 1983) confirms that the total population of moufflons in their original range is now well over a thousand animals (probably intermediate between 1,100 and 1,600), and it is slightly but constantly increasing (Table 20.1). In comparison with the 1960s, when it reached its historical minimum, the species appears to have benefited by a somewhat better oriented public interest, and by a more effective legal protection which now makes possible heavy penalties for poachers. Moreover, it could profit by a reduction in the human presence within its natural range, as there

Table 20.1: Present Status of the Moufflon in its Natural Range

Areas	Sources		
	Schenk (1976)	Schenk (1980)	Ruiu & Congiu *i.l.* (1983)
Mt. Albo	± 20	20-30	110-150
Supramonte, northern part (Corrasi-Lanaitto-Mt. Oddeu-Donanìgoro)	60-80	80-120	120-160
Supramonte, central part (P. Solitta-Gorroppu-Mt. Unnoro-Mt. Su Biu)	100-150	150	160-200
Supramonte, southern part (Mandracaia-Correboi-Mt. Genziana-Mt. Pisaneddu)	100-150	250-350	150-200
Supramonte Dorgali/Baunei (Codula di Luna-Codula Sisine-Bacu Arala)	30-50	50-60	30-60
Mt. Fenalbu/Mt. Orosei	30-50	40-60	
Mt. Gennargentu (Fonni-Desulo-Funtana Congiada-Villagrande)	150-230	210-230	260-320
Mountains south of Flumendosa River (Cuccuru 'e Mufloni-Perda Liana-Montarbu-Mt. Perdedu-Mt. Armidda)	220-310	300-360	170-270
Intermediate areas, trespassing herds, etc.	15-30		150-230?
Total	725-1,070	1,100-1,360	1,150-1,590

is now much less sheep-farming in the area than formerly. Groups of 30 to 80 moufflons are now frequently sighted in the best areas (Schenk, 1980; Ruiu, 1983). Poaching is still heavy (and only rarely discovered and punished), but births probably exceed the losses (Ruiu & Congiu, *in litt.* 1983).

Conservation and Management

Since numbers seem to have recovered quite well, the Sardinian moufflon is not so endangered at present in its natural range as it probably was only a decade ago. However, poaching still constitutes a severe threat, because of insufficient surveillance and inadequate protection of the whole territory. A yearly loss of 50-100 moufflons in 1975/1980 was considered only a very approximate guess by Schenk (1980), based both on official and confidential information. The largest incidence of poaching is reported for the Oliena and Fonni areas (where poachers even form sometimes well-organised teams), but only six fines (for at least 18 moufflons killed in all) appear to have been imposed during the last three years (Ruiu & Congiu, *in litt.* 1983).

Moreover, the natural habitat of the moufflon continues to be severely modified by various human activities. Many new roads have been built, even in the most delicate mountain areas, for timber harvesting, reafforestation, shepherd facilities, expansion of tourism, and for winter ski-resorts. These, again, facilitate poaching and disturbance. Fires represent another important cause of habitat destruction.

No conservation measures have been taken so far to protect the area. Apart from the two Regional Forests of 'Montarbu' and 'Montes' (totalling about 7,300 ha), where hunting is forbidden and a certain surveillance is guaranteed, no further steps have been made for the creation of the long-awaited Gennargentu National Park. The creation of this Park would represent the best solution for reconciling tourism and conservation, local economical activities and proper management of the natural range of the moufflon. Unfortunately many local administrators and inhabitants still oppose this project, and neither the State nor the Region have decided, so far, to create the National (or Regional) Park.

In a region like Sardinia, which attracts tourists mainly

because of its beautiful and natural landscape, the Gennargentu/Supramonte massif constitutes one of the very last Italian wilderness areas, which should be carefully kept and properly managed for the enjoyment of visitors and the benefit of local economy. The moufflon, being the most attractive element of its valuable wildlife, could become a major factor in such a policy. If well protected and undisturbed, it could easily increase its numbers and reduce the flight distance, and its presence would be much appreciated by visitors and tourists. Hopefully, when the Park becomes a reality, these few simple measures — such as the organisation of guided excursions on special nature trails, the use of some sheepfolds ('cuiles') as rustic huts for resting and relaxation, and the arrangement of salt licks or small rain-water reservoirs which moufflons would easily learn to visit — could make the search for these fine animals an exciting 'adventure' for future visitors, thus bringing important benefits to the economy and occupation of the many inland villages in the whole area.

Re-introductions and Introductions

When numbers have become sufficiently large within the moufflon natural range, re-introductions ought to be made into other suitable areas of Sardinia. The Limbara Massif, the Nieddu and Senes Mountains, the hills around Alà dei Sardi and Buddusò, the Mt. Linas/Marganai complex in the Iglesiente, appear to be especially deserving of such a re-introduction. In view of this, these areas ought to be declared and managed as regional parks or nature reserves.

Moufflons are very robust and adaptable, and re-introductions could probably be highly successful, thus extending viable moufflon populations to other areas as well, where a suitable game management could eventually be developed for hunting purposes. In short, the Sardinian moufflon appears to be worthy of an improved management, rather than to be simply abandoned to its fate within its rapidly changing natural environment.

Introductions have to be regarded differently. In fact, scattering moufflons into many different environments of the world, as has been done for many decades, is simply a biologically senseless operation. Nevertheless, in particular situations, where the presence of moufflons may enrich a depleted environment by restoring a valuable wildlife species in order to improve the conservation and management of the whole area — e.g. several different European areas, such as the Caroux in Southern France (Pfeffer & Genest, 1969) — the introduction of the moufflon can be regarded as a positive attempt at managerial improvement of nature through a correct exploitation of the aesthetic and cultural values of this ungulate.

Introductions, on the other hand, ought always to be firmly avoided when, in the releasing area, other ungulates occur which may be adversely affected by the presence of the moufflon. Competition for food and natural range is reported as being particularly heavy in chamois, especially in winter, when the nutritional resources are limited (Pfeffer & Settimo, 1973). Similar conclusions could be inferred from Gonzalez (1985), who found that the spatial overlapping of moufflons and chamois never exceeded 10-18 per cent of the total home range of both species in the Carlit Massif, Pyrenees. Overlap was minimal during the winter, with a marked altitudinal segregation, due to reciprocal intolerance, even in the areas of overlapping. Filej (1976), who studied the moufflon/chamois relations in the Urslja Gora region, also found that the introduction of 12 moufflons (rapidly increased up to 80, from 1966 to 1973) displaced the chamois from their natural range, greatly reduced their density and produced negative alterations in their state of health. Since moufflons are dominant over chamois, the latter tend generally to be pushed to marginal habitats where food is scarcer and where they are more exposed to poaching. The presence of moufflons, in particular, represents by itself a major threat for the very localised and endangered Chartreuse chamois (*Rupicapra rupicapra cartusiana* Couturier 1938), for whose conservation the extirpation of moufflons has been proposed as an urgent need (ECPS, 1982).

References

Cassola, F. (1976) 'Il Muflone di Sardegna: Importanza, Stato Attuale e Problemi di Conservazione' in *S.O.S. Fauna. Animali in Pericolo in Italia*, WWF, Camerino, pp. 67-107

Demeautis, G. (in press) 'First Facts in the Study of the Wild Sheep (*Ovis ammon musimon*) in the Frame of the Corsican Regional Natural Park' in *Abst. Inter. Conf. Terrestrial and Freshwater Vertebr. from Mediterranean Islands*, Evisa 1

Dubray, D. (in press) 'Abundance, Population Structure and Dynamics of the Corsican Wild Sheep (*Ovis ammon musimon* S.) in the Eastern Sector of the Cinto Massif (Haute-Corse), and Analysis of the Protection Role of the National Hunting Office Reserve of Asco' in *Abst. Inter. Conf. Terrestrial and Freshwater Vertebr. from Mediterranean Islands*, Evisa 1

ECPS (1982) '*Rupicapra rupicapra cartusiana* Couturier 1938' in *A Draft Community List of Threatened Species of Wild Fauna and Flora*, 2 vols, ECPS, Commission of the European Communities, (Mimeographed), pp. 1-4

Filej, J. (1976) 'Influence des Colonies de Mouflon sur la Population de Chamois dans l'Urslja Gora', *Proc. II Symposium Chamois*, Bled 1976. (Mimeographed), pp. 1-6

Gonzalez, G. (1985) 'Seasonal Fluctuations in the Spatial Distribution of Chamois and Moufflons on the Carlit Massif, Pyrenees' in S. Lovari (ed.), *The Biology and Management of Mountain Ungulates*, Croom Helm, London, pp. 117-23

Massoli Novelli, R. (1973) 'Sardegna amara per Cervi e Mufloni', *Il Cacciatore Italiano*, *13*, 19-20

Perco, F. (1981) 'Muflone, *Ovis musimon* Pallas 1811' in C.N.R., *Distribuzione e Biologia di 22 Specie di Mammiferi in Italia.* Roma, pp. 155-9

Pfeffer, P. (1967) 'Le Mouflon de Corse (*Ovis ammon musimon* Schreber, 1782). Position Systématique, Écologie et Éthologie Comparées', *Mammalia*, *31*, Suppl., 1-262

Pfeffer, P. & Genest, H. (1969) 'Biologie Comparée d'une Population de Mouflons de Corse (*Ovis ammon musimon*) du Parc Naturel du Caroux', *Mammalia*, *33*, 165-92

Pfeffer, P. & Settimo, R. (1973) 'Déplacements Saisonniers et Competition Vitale Entre Mouflons, Chamois et Bouquetins dans la Reserve du Mercantour (Alpes Maritimes)', *Mammalia*, *37*, 203-19

Reinhardt, S. & Schenk, H. (1969) 'Zur Verbreitung der Gross-säuger auf Sardinien', *Bonner Zool. Beitr.*, *4*, 429-36

Ruiu, D. (1983) *Baronie a Piedi*, Comunità Montana Baronie, Torpè, pp. 1-56

Schenk, H. (1972, 1975) 'Ambiente Faunistico. Per il Sistema dei Parchi e per il Limbara' in A. Lacava (ed.), *Studio del Parco del Limbara nel sistema regionale dei parchi. Prima e seconda fase*. Centro Regionale di Programmazione, Cagliari

—— (1976) 'Analisi della Situazione Faunistica in Sardegna. Uccelli e Mammiferi' in *S.O.S. Fauna, Animali in Pericolo in Italia*. WWF, Camerino, pp. 465-556

—— (1980) 'Preliminary Report on the Status of the Moufflon (*Ovis musimon*) in Sardinia', (Mimeographed), pp. 1-7

Silvestri, A. (1974) 'Indagine sulla Consistenza del Muflone (*Ovis ammon musimon*) in Italia', *Natura*, *65*, 134-42

Van Haaften, J.L. (1973) 'The Moufflon of the Mediterranean region', (Mimeographed), pp. 1-4

CHAPTER 21

Distribution and Status of the Spanish Ibex (*Capra pyrenaica* Schinz)

CONCEPCIÓN L. ALADOS
Estaciòn Experimental de Zonas Aridas, 1 General Segura Str., Almería (Spain)

Abstract

The status and distribution of the Spanish ibex (*Capra pyrenaica*) is as follows. Of the four recognised subspecies, *lusitanica* has become extinct, *pyrenaica* amounts to only 25-30 individuals, whereas the other two (*victoriae* and *hispanica*) are well represented in the central and eastern areas of the Iberian Peninsula.

Introduction

In the Pliocene age (7 million years ago) various forms of Central Asian caprids reached Europe. At that time the genera *Capra* and *Ovis* were already clearly differentiated and it is not until the Riss glaciation that the true Pyrenean form of ibex is found, clearly distinct from the Alpine form by Würm II (Gonzalez, 1982). Reacting to the cold phases of the Middle Pleistocene, the ibex spread to lower altitudes over all Europe: its remains are found in the Iberian Peninsula (Gibraltar), France (Mentòn), Germany (Turingia), Italy (Puglia) and the Balcanic Peninsula. During the following warmer periods, however, the ibex moved back again to higher altitudes, keeping to some mountain ranges, where they can still be found.

Numerous fossil remains and prehistoric rock paintings of the Spanish ibex show that the species abounded in the Iberian

Peninsula during the Palaeolithic and Neolithic periods.

With time, the ibex populations decreased considerably and even disappeared from certain areas. By the mid-nineteenth century they were extinct in the French part of the Pyrenees and, around 1890, in the NW of the Iberian Peninsula. By the end of the last century the only known colony in the whole Pyrenees was on Monte Perdido, in Ordesa valley (Huesca County).

Only in 1905 were conservation measures undertaken to preserve this species and, under the aegis of King Alfonse the XIIIth, the National Refuge of Sierra de Gredos was established, saving a very small population of ibex (12 animals). Not much was done to protect this caprid until after the Civil War (about 1950), when an effective programme for the conservation of ibex was started.

The evaluation of the status and distribution of *Capra pyrenaica* reported in this paper is based on data from official reports or scientific and game journals: Galindo (1965), Rodrìguez (1969), Cerda & Peña (1970), Labarere (1971), Engländer (1972), Ortuño & Peña (1979), Gonzalez (1982), Pascual (in press).

Taxonomy

According to Cabrera (1914) and mainly on the basis of horn shape and coat colour, four subspecies of *Capra pyrenaica* existed in the Iberian Peninsula.

1. *C.p.pyrenaica* Schinz 1838. Type Locality: Spanish Pyrenees. Today the species can be found on Monte Perdido (Ordesa) only.
2. *C.p. victoriae* Cabrera 1911. Type Locality: Madrigal de la Vera. At present only on Sierra de Gredos.
3. *C.p.hispanica* Schimper 1848. Type Locality: Summits of Veleta and Mulhacen, Sierra Nevada. Today present in the area from Gibraltar to the mouth of the Ebro river, including Sierra Morena as well.
4. *C.p.lusitanica* Schlegel 1872. Type Locality: Sierra de Geres, in Portugal. Extinct in 1890.

Habitat and Status of the Populations (Figure 21.1)

Maestrazgo

The National Refuge of Puertos de Tortosa y Beceite (30,418 ha) was established in 1966. It is located on the eastern end of the Iberian Mountain system. The highest altitude is 1,447 m a.s.l. The structure is calcareous and the soil is affected by intense erosion. The climate is Mediterranean temperate, mean rainfall ranging from 480 mm annually at the lowest altitudes to 1,000 mm at the highest. Mean temperature is 12.4°C on Monte Fou, varying from −7°C in winter to 40°C in summer.

The predominant plant species at the lower altitudes are *Pinus halepensis*, shrubs of mainly *Quercus ilex*, *Q. coccifera*, *Pistacia lentiscus* and *Crataegus monogyna*. From 600-700 m a.s.l. upwards the dominant species is *Pinus nigra* whereas above 1,000-1,200 m it is *Pinus sylvestris*.

The evolution of the ibex population of the National Refuge is presented in Table 21.1 (Ortuño & Peña, 1979).

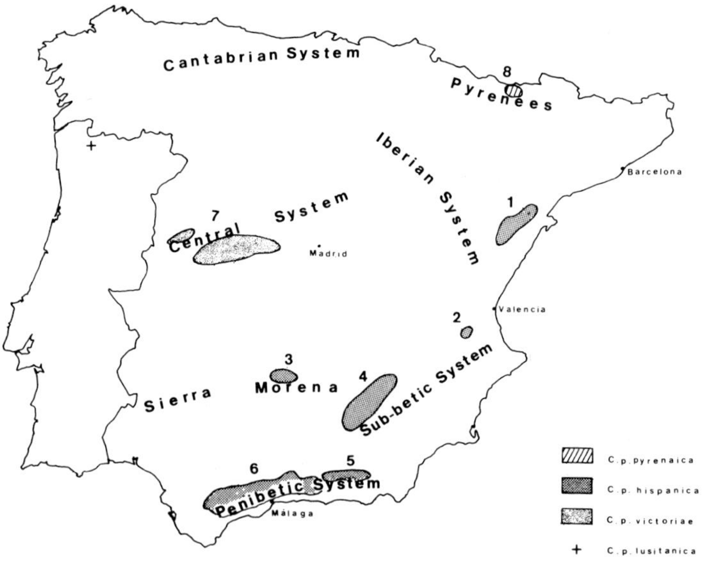

Figure 21.1: Distribution of the Populations of Spanish Ibex. 1, Maestrazgo; 2, Muela de Cortes; 3, Sierra Madrona; 4. Sierras de Cazorla y Segura; 5. Sierra Nevada; 6. Ronda and nearby ridges; 7. Sierra de Gredos; 8, Valle de Ordesa. The population of Sierra de Geres in Portugal became extinct in 1890 (+)

Table 21.1: Evolution of the Maestrazgo Population

Year	1966	1970	1974	1977	1979
Population size (no. ind.)	450	675	2,160	3,190	3,800

The size of the population in 1966 and 1970 was probably underestimated.

The total, including those in the National Refuge and the surrounding area, was 875 animals in 1970 and 4,320 in 1977 (Peña, 1978).

The last official census, in 1981, recorded 3,900 ibex in the National Refuge, 800 in private game reserves and 400 in other wildlife areas.

Muela de Cortes (Valencia District)

The National Refuge was created in 1973 to protect a small remaining population of ibex and 57 individuals from Sierra de Cazorla were also released to increase the local population.

The reserve is located on the periphery of the Iberian Mountain System and it extends for 36,009 ha. This calcareous, steep outcrop includes altitudes ranging from 350 to 1,016 m a.s.l. The climate is Mediterranean temperate, with an annual mean temperature of 14.4°C; the annual rainfall is 601 mm.

The predominant plant species, up to 750 m of altitude, are *P. halepensis* and *Q. ilex*. Higher up, *P. nigra* is the most abundant pine. In the most humid areas *Q. coccifera*, *Juniperus sabina*, *J. communis* and *Arbutus unedo* are abundant, whereas in the driest zones *Erica*, *Cistus* and *Rosmarinus* are common.

In 1978 the ibex population numbered 460 individuals in the reserve (Ortuño & Peña, 1979), and totalled 600 head for the reserve plus the surrounding area (Peña, 1978).

Sierra Madrona

This is located on Sierra Morena, Ciudad Real province. Some ibex were introduced 10 years ago and have produced the population now found mainly in private game preserves (90 per cent) over an area of 20,000 ha; numbers, according to official reports, are about 25-30 animals.

Sierras de Cazorla y Segura

This is located at the Sub-betic Mountain System. The size of the National Refuge is 66,367 ha, with calcareous peaks from 600 to 2,170 m a.s.l. At Navas de San Pedro (1,400 m) the annual mean temperature is 8°C; the mean annual rainfall is 1,500 mm.

Three species of pines are found in the conifer forests: *P. halepensis* occurs between 600 and 850 m of altitude, associated with *Q. ilex*. The *P. pinaster* forest, found between 850 and 1,200 m high, is mixed with *Q. ilex*, *Q. faginea* and *Acer granatensis*. The *P. nigra* wood is found between 1,200 and 1,850 m of altitude.

Meadows occur at higher altitudes.

When the National Refuge was established in 1960 the population totalled about 300 animals, but later figures provided by Peña (1978) indicate 8,000 animals in the National Refuge and 9,200 on the whole Cazorla and Segura range; recent information (1983) shows about 6,000 individuals for the Reserve.

Sierra Nevada

This mountain system includes the highest altitudes in the whole peninsula (3,366 m at La Alcazaba, the highest summit). The highest, central part of this system is fairly flat and rounded, schist being its geological structure. The lower and steeper peripheral zones are made up mainly of limestone and sandstone.

The National Reserve, located in the central part of this system, extends for 35,430 ha.

The climate is milder on the southern slopes of the system. Winter snowfalls are common and the snow cover lasts from the autumn to the spring.

Over 2,900 m a.s.l. lichens and several meadows of *Festuca clementei*, *Agrostis nevadensis* and *Poa laxa* occur. In the tundra of Sierra Nevada (2,200-2,900 m) the dominant shrub species are *J. sabina*, *J. communis* and *Cytisus scoparius*. At lower altitudes there are deciduous forests of *Castanea sativa*, *Celtis australis* and *Acer* spp; *Pinus nigra* and *P. halepensis* are found in a few areas.

The population of ibex was dwindling in 1936, and protection was actually started in 1960. The evolution of the size of the ibex population in the Reserve is shown in Table 21.2.

Total size of the Sierra Nevada population in 1970 was 790 animals; 2,000 individuals in 1977 (Peña, 1978); 2,700 individuals in 1982 (National Reserve only).

Table 21.2: Evolution of the Sierra Nevada Population

Year	1964	1975	1978
Population size (no. ind.)	400	624	945

Ronda and Nearby Chains

These mountain chains belong to the Penibetic system and extend from Sierra Bermeja (Malaga district) to Sierra de Lujar (Granada district). Sierra de Ronda is the most important ibex area. The terrain is calcareous and the highest altitude reaches over 2,000 m.

In spite of intense rainfall these are dry ridges, mainly because of the severe summer droughts.

Two National Reserves are located in this area: Refuge of Sierra de Tejeda y Almijara (20,398 ha) and Refuge of Serranìa de Ronda (23,663 ha).

The principal plant species found are *Abies pinsapo*, *J. sabina*, *J. communis*, *P. nigra*, *P. halepensis*, *Q. ilex*, *Q. faginea*, *C. monogyna*, *Ceratonia siliqua*, *Sorbus* and *Acer*.

The evolution of population size in the two ibex refuges is presented in Table 21.3 (Ortuño & Peña, 1979). For the whole population from Sierra Bermeja to Sierra de Lùjar, 1,240 ibex were counted in 1970, the numbers increasing to 2,685 in 1977 (Peña, 1978).

Table 21.3: Evolution of the Population of Ronda and Nearby Chains

National Reserve of Sierras de Tejeda y Almijara							
Year	1970	1973	1978				
Population size (no. ind.)	75	200	480				
National Reserve of Serranía de Ronda							
Year	1958	1960	1961	1965	1971	1973	1978
Population size (no. ind.)	204	320	414	785	1,105	1,560	1,450

Sierra de Gredos

This mountain chain is a part of the Central Orogenic System. Almost all its central area including the highest altitudes is within the National Reserve (27,222 ha). Terrain structure is mainly granite.

Low temperatures reach −20°C in winter; the snow cover continues for a great part of the year and summer temperature is mild (19°C)

The vegetation is made up mainly of *Quercus toza*, *P. pinaster*, *P. sylvestris* and *Erica arborea*.

The evolution of the Reserve population is shown in Table 21.4 (Ortuño & Peña, 1979).

Table 21.4: Evolution of the Sierra de Gredos Population

Year	1905	1928	1932	1949	1952	1955	1958
Population size (no. ind.)	12	1,200	600	1,915	1,747	2,032	1,827
Year	1962	1965	1969	1974	1976		
Population size (no. ind.)	1,923	2,289	3,217	3,611	3,913		

The Sierra de Gredos population of 3,580 ibex in 1970 and 9,320 in 1977 (Peña, 1978) made up the whole Reserve population.

The last official census, in 1983, recorded 3,145 ibex in the National Refuge.

Valle de Ordesa

Located in the Aragòn district of the Pyrenees. The ground is calcareous, from 1,000 to 2,300 m a.s.l. Average minimum temperature (at 1,880 m) is −6°C. Snow lasts unmelted throughout the whole winter.

The vegetation is typical of the montane (main species are *P. sylvestris*, *Buxus sempervirens*, *Echinospartum horridum*, *Abies alba* and *Fagus sylvatica*) and subalpine altitudes (mainly *Pinus uncinata*, *Rhododendron ferrugineum* and *Vaccinium myrtillus*).

In 1907 Couturier & Cabrera gave a total number of 8 or 9

ibex only, whereas Gourdon counted a total of 40 animals (Pascual, in press). In 1970 Peña (1978) reported only 12 animals, increasing to 20-25 in 1977. Finally, in the winter of 1980-81, Pascual (in press) counted 30 animals: 4 adult males, 1 juvenile male, 4 young, 6 adult females and 15 individuals whose age and sex classes could not be determined.

By adding together all the populations of the Spanish ibex, we have a total number of 28,190 animals, of which 19,410 live in national game reserves and 1,970 in other wildlife areas (Peña, 1978).

The survival of *C. pyrenaica* is no longer in danger, but two points of utmost importance need mentioning.

The three most important ibex populations (those in Cazorla and Segura, Gredos and Tortosa y Beceite) either live near human centres of high population density or are visited regularly by a great number of tourists. The disturbance to those ibex colonies is great and very apparent to anyone; it includes expulsion from regular feeding areas, molesting and sometimes even separating mothers from newborn young.

The common practice of mixing animals from different populations may disrupt the natural situation and should therefore either be done with much care, or not at all.

References

Cabrera, A. (1914) 'Mamîferos' in *Fauna Iberica*, 320, Madrid

Cerda, J.M. & Peña, J. (1970) *La Cabra Montés Espanola*, Congreso Internacional de la Caza, Madrid

Engländer, H. (1972) 'Steinbocke in Spanien', *Verein zum Schutze der Alpensplanzen und -Tiere*, *37*, 1-8

Galindo, F. (1965) *La Capra Pyrenaica Hispanica de los Puertos de Beceite (Teruel)*, Imprenta Escuela Provincial. Teruel

Gonzalez, G. (1982) 'Eco-Ethologie du Bouquetin en Sierra de Gredos', *Acta Biologica Montana*, *1*, 177-215

Labarere, J. (1971) 'Le Bouquetin des Pyrénées', *Pyrénées*, *87*, 163-17

Ortuño, F. & Peña, J. (1979) *Reserves y Cotos Nacionales de Caza*, Ed. INCAFO. Madrid

Pascual, R. (in press) 'El "bucardo" (*Capra pyrenaica pyrenaica* Cabrera 1911) en el Parque Nacional del Valle de Ordesa', *XV Congreso Internacional de Fauna Cinegética y Silvestre*. Trujillo (Càceres) Espana, 1981

Peña, J. (1978) *La Cabra Montés en Espana*, Congreso Internacional de la Caza, Suiza

Rodríguez, M. (1969) *La Cabra Montes en Sierra Nevada*, Documentos tecnicos. Serie cinegètica 1. Ministerio de Agricultura

CHAPTER 22

Harvesting an Atypical Ibex (*Capra ibex* L.) Population: A Management Plan

H. KOFLER and W. SCHRÖDER

Institute of Wildlife Research, University of Munich, Amalienstrasse 52, 8000 München 40 (FRG)

In 1960 about 20 ibex were introduced into the Hochlantsch mountains, about 30 km north of Graz, Austria. This small massif is the easternmost mountainous area of the Alps; further east the Alps ebb out in forested hills. The area is of limited size (about 2,000 ha), at a very low altitude for the ibex (1,722 m). Artificially created alpine pastureland has partly replaced the forest, which formerly covered the entire massif except for the steep limestone rocks.

The ibex have increased to over 300 animals in less than 20 years (Figure 22.1) reaching the geographic limit of this small habitat (Figure 22.2). Over the same period the chamois have declined. Severe competition for food was concluded as being the factor responsible for the chamois decline: the numerous ibex, having a lower food quality requirement, may have exerted such a heavy pressure on the vegetation that the quality of food available for chamois declined (Schröder & Kofler, in press). Severe competition between these mountain ungulates may have occurred in this area because of the mild winters which did not restrict the ibex to small wintering areas as in high altitude populations. Thus, range overlap of ibex and chamois in the critical, cold season enhanced their competition.

The harvesting of low numbers of ibex began in 1965. In 1980 the 10 independent hunting districts grouped together to form a single Management Unit over all the ibex range. The following

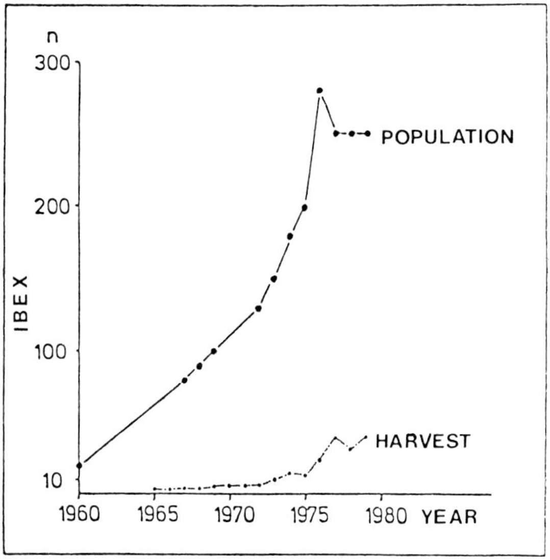

Figure 22.1: Increase of Ibex Following the Introduction, According to Annual Counts. In 1965 cautious harvesting began

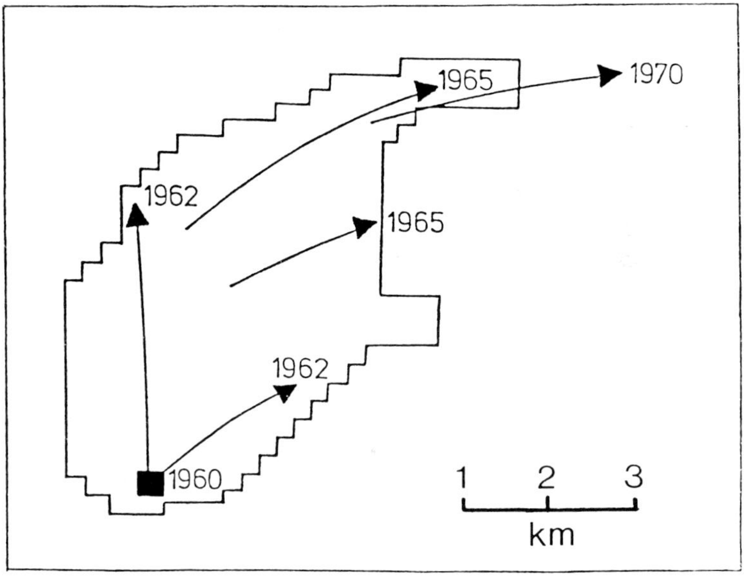

Figure 22.2: The Dispersal of Ibex from the Release Site. The main ibex range is outlined

management objectives were decided: (a) to prevent forest damage; (b) to maintain both species, chamois and ibex; (c) to harvest the ibex.

The Management Unit agreed to count ibex and chamois twice a year to document the population trend and to provide sound information to plan the harvest (Kofler, 1983).

Most of the previous data on ibex harvesting came from high altitude mountain areas, thus being hardly applicable to this rapidly growing, low altitude population. It was planned to build a computer model and to use simulations to determine the harvest plan (Schröder, 1983).

Signs of an eruptive population up-swing with a threat of overshoot and the competition with chamois were already detectable, although the early harvesting may have prevented a too severe overshoot. Caughley (1976) has developed the model of an ungulate population which can be regulated, if harvesting begins when the population is about halfway along its up-swing. Caughley's work indicates that the final equilibrium between vegetation and animal populations may thus be reached without dramatic oscillations and the harvest can be progressively reduced. Our goal, however, is to achieve a target-population size well below such a theoretical carrying capacity, through sustained yield harvesting (Figure 22.3). In this way we hope to

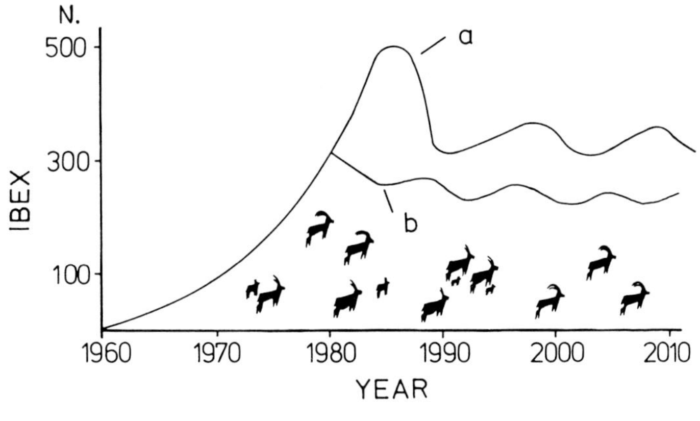

Figure 22.3: (a) Expected Overshoot and Equilibrium of Ibex Population Without Harvesting; (b) Target Population Maintained through Sustained Yield Harvesting

find an 'optimum' population in accordance with the management objectives mentioned above.

References

Caughley, G. (1976) 'Wildlife Management and the Dynamics of Ungulate Populations', in T.H. Coaker (ed.), *Applied Biology*, Academic Press, New York, pp. 101-93

Kofler, H. (ed.) (1983) *Der Steinbock: Ansprüche, Einburgerung, Bejagung*. Tagungsbericht Wildbiologische Gesellschaft, München

Schröder, W. (1983) 'Zur jagdlichen Planung in Steinbockpopulationen' in H. Kofler (ed.), *Der Steinbock*, Wildbiologische Gesellschaft, München, pp. 37-49

Schröder, W. & Kofler, H. (In press). 'Coexistence and Competitive Exclusion between Ibex and Chamois'. *Acta Zool. Fenn.*

CHAPTER 23

The Impact of Tourism on Chamois Feeding Activities in an Area of the Abruzzo National Park, Italy

A. CEDERNA and S. LOVARI
Istituto di Zoologia, Strada dell'Università 12, 43100 Parma (Italy)

Abstract

A positive correlation was found between the number of visitors and that of disturbers (p=0.05) throughout the study duration. The peak of the tourist pressure and harassment events occurred in mid-summer. Hikers leaving the main path determined the large majority of disturbances and those heading straight to the chamois underlay most harassment events and flight reactions. A tangential approach proved significantly less disturbing (p<0.001). Repeated harassment of the same herd or individual was common. Up to nine successive chamois pursuits were recorded. Usually the same tourist insisted on consecutive harassing attempts. The most common purpose of disturbers was to take close-up pictures of chamois. Flight distance was inversely correlated to chamois group size (p≤0.05) and fleeing individuals preferred to move to an equally steep (56.48 per cent) or a steeper (38.07 per cent) escape terrain (N=239). Consecutive harassment events and number of chamois retreating to rocky outcrops were significantly correlated (p≤0.05). Grazing was the activity most affected by tourist harassment. Ruminating bouts were interrupted less frequently. The fairly regular rhythms of undisturbed grazing and ruminating chamois were strongly affected by human interference especially in the morning.

Introduction

Published material on research carried out to assess qualitatively and quantitatively the direct impact of tourism on ungulate species in protected areas is scanty. Hicks & Elder (1979) found that distance, juxtaposition, age, sex composition and size of the animal herd are important factors in bighorn (*Ovis canadensis*) reactions to humans. They also stated that bighorn-human encounters were limited to specific locations and were not adversely affecting the bighorn population as a whole. Furthermore, MacArthur *et al.* (1982) showed that cardiac and behavioural responses of these sheep to an approaching human are the greatest if the visitor appears from over a ridge or he is with a dog. This paper is a preliminary report of our research evaluating the nature of tourist-chamois interactions in an attempt to unravel the relevant effects on some chamois habits. Such data may eventually be used to support the enforcement of restrictive management measures on tourism in areas where conflicts between tourist and chamois activities occur.

The Abruzzo National Park territory supports the last, remnant population of 350-400 Apennine, or Abruzzo, chamois (*Rupicapra rupicapra ornata* Neumann 1899) (Lovari, 1977). Such a rare subspecies has been recorded as 'vulnerable' by IUCN (Thornback, 1980) and special care must be taken to preserve it.

This research could not have been carried out without the collaboration and support of the Abruzzo National Park director, F. Tassi, to whom we are very grateful. Also, we should be indebted to those undisciplined tourists who, unknowingly, provided data for our study.

Material and Methods

Observations were carried out on 10-20 July, 10-20 August, 1-11 October 1981, and 9-19 June 1982, continuously from 07.00 hours to 16.00 hours. In October we recorded no data till 08.00 hours, as the shortened day-length determined a later arrival of the chamois on the alpine meadows from their rocky, precipitous night dormitories. Data were noted on two check-sheet models (cf. Hinde, 1973): one was utilised for tourist-chamois interactions; the other, filled in at 15-minute intervals, was used for the chamois activity rhythms, in October 1981 and June 1982. Manual notes allowed the recording of additional data not considered in the check-sheet models. The chamois were watched from a vantage point, usually at a distance of about 40-60 metres or further, with the observer as much as possible not in sight of the animals. In fact, previous data had shown that the alert distance of the chamois herds under study ranged around 25-30 metres and that the downwind or upwind position of the visitor did not bear any significant importance for the chamois reactions (Cederna & Lovari, 1983).

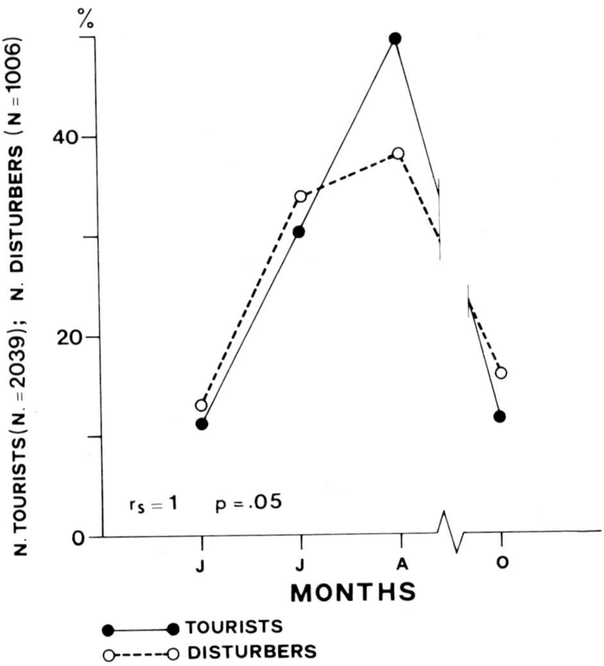

Figure 23.1: Relationship between Number of Visitors and Number of Chamois Disturbers during the Tourist Season

Percentages in Figure 23.1, within the 'Head to' histogram and between the 'Head to' and 'Pass by' histograms, were respectively checked through the test to evaluate the actual percentage versus a fixed percentage level, i.e. 50 per cent, and through the test to evaluate the significance of difference between percentages in large samples reported by Pollard (1977: 138 and 144, respectively). Unless otherwise stated, correlations were tested by parametric correlation coefficients (cf. Pollard, 1977).

Study Area

Data were collected in the upper Val di Rose (1,750-1,942 m a.s.l.), a valley of the Abruzzo National Park, Italy, in the heart of the chamois range. This is an amphitheatre above the timberline, where the chamois density is the greatest (about 38 ind./km^2) outside the winter months (Lovari, 1984). Limestone ridges and rocky outcrops make up about 40 per cent of the amphitheatre, while Alpine meadows cover the remaining part. These lie mainly on the left and the lower right slopes. There is a central path stretching from the lower to the upper part of the valley. Two smaller trails lead respectively to the left and right ridges, departing from the upper section of the amphitheatre bottom. Only the main path is open to tourists by the Park regulation, which also forbids camping, loud noises and any direct harassment of wildlife. However, fines are negligible.

Results

Of the visitors (N = 2039) 49.33 per cent interfered, singly or in group, with chamois activities. There was a significantly direct correlation between the total number of tourists and that of the disturbers throughout all the period of our observations ($r_S = 1$; $p = 0.05$, Spearman Rank Correlation Coefficient) (Figure 23.1). The peak of the tourist pressure occurred in July and particularly in August (70.24 per cent of all disturbance events; N = 289). Only 29.33 per cent of the disturbances (N = 300) were caused by tourists on the path, who interfered with chamois which were grazing or lying down in the trail vicinity. In fact, the most harassing occurrences (70.77 per cent) were caused by hikers who had left the path. Tourists heading straight towards the chamois disturbed them 69.12 per cent of the times (N = 298) and determined 77.96 per cent of flight reactions (N = 236). This type of approach was the main factor causing flight, whereas the 'passing-by' tourist was responsible for no more than 57.14 per cent of occurrences (Figure 23.2). Out of 294 active disturbance events, a second harassment was recorded 49.65 per cent times, and additional ones occurred in 30.61 per cent of the cases. Up to nine successive chases were observed for the same chamois group. Usually (81.50 per cent; N = 146) the same disturber insisted in such repeated harassing attempts.

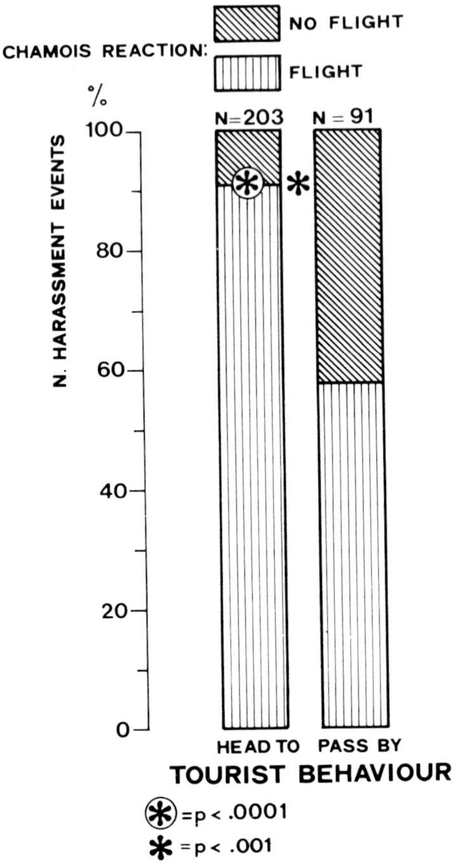

Figure 23.2: Occurrence of Chamois Flight Reaction to Tourists Heading Straight to or Passing by Them

Tourists trying to get close-up pictures of the chamois were responsible for over half of the disturbance events (Figure 23.3). When further chases occurred, this percentage increased sharply: up to 90 per cent (Figure 23.3).

Flight reactions were inversely correlated to chamois group size ($r = -0.99$; $p \leqslant 0.05$). Fleeing individuals selected an equally steep terrain in 56.48 per cent of the cases and a steeper one in 38.07 per cent (N = 239): the latter preference increased (42.96 per cent; N = 128) in case of further harassments. Alert

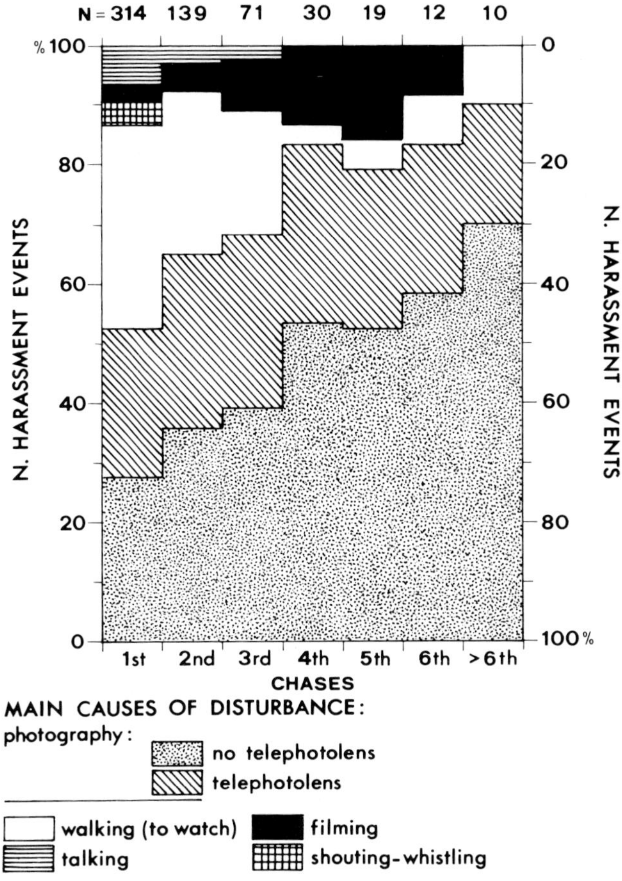

Figure 23.3: Harassment Events in Relation to Main Causes of Disturbance by Tourists on Repeated, Consecutive Approaches to Chamois

and flight distances were initially greater in flocks containing one-third, or more, of kids in comparison with those of groups formed by a relatively smaller number of kids. Such distances tended to coincide slowly from early summer to autumn. A significant direct correlation was found between the number of consecutive disturbance events and that of chamois retreating to rocky outcrops ($r = 0.989$; $p \leqslant 0.05$).

Grazing was the activity most affected by tourist harassment (67.66 per cent; N = 300) whereas ruminating bouts were

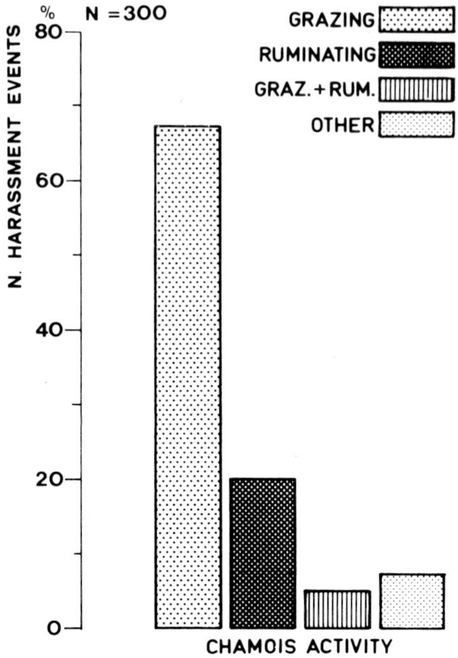

Figure 23.4: Chamois Activities Interrupted by Tourist Harassment

interrupted less frequently (20 per cent) (Figure 23.4). Figures 23.5 and 23.6 show respectively the extent of the influence of harassment on grazing and ruminating, in June. Sharp peaks and falls impaired the regular rhythms of these important nutritional activities of the chamois. Data for October (Cederna, in preparation) provided similar figures.

Discussion

One may expect that chamois in larger flocks are less shy than in smaller groups or alone, because of the calming-down effect of conspecifics and the relative security that may be derived from being in a flock. Thus, our data on flight distance are consistent with the information reported by Altmann (1958) on *Alces alces* and *Cervus canadensis*, by Walther (1969) on *Gazella thomsoni*, by Hicks & Elder (1979) and MacArthur *et al.* (1982) on *Ovis*

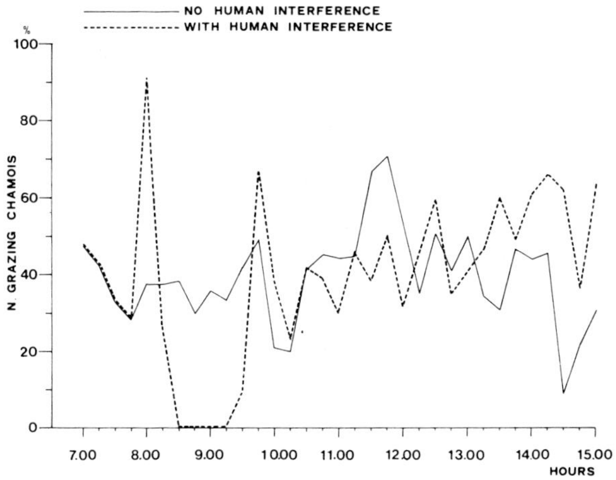

Figure 23.5: Daily Grazing Rhythms of Chamois in Days With or Without Interference by Tourists. June data

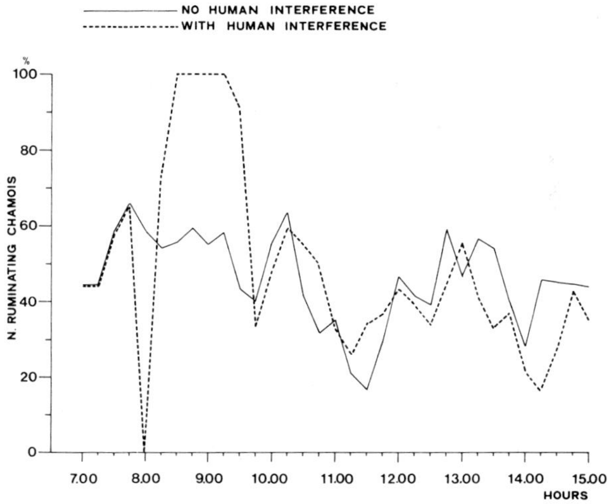

Figure 23.6: Daily Ruminating Rhythms of Chamois in Days With or Without Interference by Tourists. June data

canadensis, and by Lipetz & Bekoff (1982) on *Antilocapra americana*. On the other hand, group composition has some influence on flight reactions: in fact, a relatively higher number of kids implies a higher number of mothers, twins being rare in chamois (e.g. Couturier, 1938) and extremely rare in the Park population (Lovari, in preparation), which in turn are easily alarmed when the kids are still very young. This may explain why the presence of a high number of kids determines greater alertness and flight distances in July and, to a lesser extent, in August but not in October, when the kids are older and relatively independent. The fairly regular rhythms of grazing and ruminating are strongly affected by human interference because, when the first tourists reach the chamois area in the early morning, they immediately go for the chamois. Then, nearly the whole flock stands up, withdraws from the disturbers and approaches the nearest rocky retreat. Often grazing is resumed while the chamois move away. That habit explains the sharp peak of grazing chamois recognisable in Figure 23.5 at 08.00 hours. As 49.65 per cent of the tourists insist on pursuing the chamois, nearly all of these reach rocky outcrops where they lie down again to ruminate, 'besieged' by the oncoming hikers till mid-morning. As the tourist pressure slackens, the chamois resume grazing. In June the impact of such a human interference is particularly heavy in the morning, which is reflected in the rhythm perturbation. Unfortunately no comparable data are available for the peak of the tourist season (July-August) but one could expect an even greater impact in connection with the much heavier tourist pressure and the relevant occurrence of harassment events.

As our data suggest that only some of the tourists interfere heavily with the chamois, restrictive measures may be aimed mainly at these undisciplined visitors. Thus, a severe, but selective, control by Park wardens on the tourists in the study area and other comparable parts of the chamois range, e.g. Mt. Amaro, is necessary and probably the only realistic measure enactable. Patrolling should occur especially from 08.00 hours to 14.00 hours, ideally from June to October. At present, wardens are on duty in the Val di Rose area from mid-July up to early September, but further encouragement — and above all control — seem to be needed to make them enforce properly the present Park regulation. Special emphasis should be given to keeping visitors to the path, leaving the meadows at the disposal of chamois as well as other wildlife. High-pitched noises (screams, whistles) make the chamois react twice as far as do the other causes of harassment (Cederna & Lovari, 1983): these disturbances may deserve a higher fine.

These data, those available in literature and others still unpublished (Cederna, in preparation; Lovari, in preparation) suggest that tourism does interfere with wildlife habits. To date it is not yet known in detail the extent of its practical effects on ungulate populations, but an effort must certainly be made to minimise its impact by enacting — and enforcing — a suitable regulation, especially in relatively small-medium size parks such as the Abruzzo National Park, where wildlife has not so many alternative retreat areas. In practice no great improvement may occur unless the Park Officials are given by the State enough personnel and an appropriate legal power. Unfortunately, this is not yet the case in Italy and other countries as well.

References

Altmann, M. (1958) 'The Flight Distance in Free-Ranging Big Game', *J. Wildl. Manage.*, *22*, 207-9

Cederna, A. & Lovari, S. (1983) 'Behavioural Interactions between Tourists and Abruzzo Chamois. Abst. 8th Meet. Ital. Soc. Anim. Behav., Pisa 1982, in *Monitore Zool. Ital.*, *17*, 184-5

Couturier, M. (1938) *Le Chamois*. Arthaud, Grenoble

Hicks, L.L. & Elder, J.M. (1979) 'Human Disturbance of Sierra Nevada Bighorn Sheep', *J. Wildl. Manage.*, *43*, 909-15

Hinde, R. (1973) 'On the Design of Check-Sheets', *Primates*, *14*, 393-406

Lipetz, V.E. & Bekoff, M. (1982) 'Group Size and Vigilance in Pronghorns', *Z. Tierpsychol.*, *58*, 203-16

Lovari, S. (1977) 'The Abruzzo Chamois', *Oryx*, *14*, 47-50

—— (1984) *Il Popolo delle Rocce*. Rizzoli, Milan

MacArthur, R.A., Geist, V. & Johnston, R.H. (1982) 'Cardiac and Behavioral Responses of Mountain Sheep to Human Disturbance', *J. Wildl. Manage.*, *46*, 351-8

Pollard, J.H. (1977) *Numerical and Statistical Techniques with Examples mainly from the Life Sciences*. Cambridge University Press, Cambridge

Thornback, J. (compiler) (1980) *A Draft Community List of Threatened Species of Wild Fauna and Flora. Mammals.* 2 vols, ECPS, Commission of the European Communities. Mimeographed

Walther, F.R. (1969) 'Flight Behaviour and Avoidance of Predators in Thomson's Gazelle (*Gazella thomsoni* Guenther 1884)', *Behaviour*, *34*, 184-221

CHAPTER 24

How to Find Suitable Ibex (*Capra ibex* L.) Habitat Using Landsat Imagery

G. WIERSEMA and W. SCHRÖDER
Wildbiologische Gesellschaft, München (West Germany)

Abstract

Analysis of Landsat imagery is used as one important step in the development of ibex habitat suitability maps. Using this method, the snowfree and vegetated winter areas, being images of those factors which best determine good ibex range, can be found. Multi-temporal images show seasonal aspects of food availability. For the Gran Paradiso and Vanoise National Parks, predicted winter habitat was found to coincide with observed winter distribution of ibex.

Introduction

Some mountain ungulates can recover well from severe losses. Others have difficulties. The Alpine ibex with its narrow habitat requirements has to be re-introduced into many distinct mountain ranges because of the limited colonising ability of this species.

In the past, many populations were introduced into unsuitable or marginal habitats. With new knowledge about the habitat requirements and limiting factors of ibex, it is now possible to develop a method for delineating good ibex habitat for further introductions (Wiersema, in press). This paper describes the application of Landsat imagery in the process of developing ibex suitability maps (Figure 24.1)

The Commission of the European Communities supported the

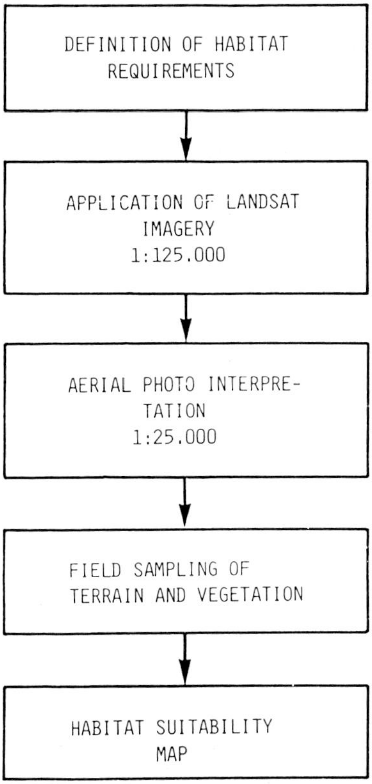

Figure 24.1: Steps in Constructing a Habitat Suitability Map

satellite application research by a Scientific and Technical Research Grant, awarded to the first author. Satellite image processing was undertaken at the ITC (International Institute for Aerial Survey and Earth Sciences, Netherlands), under the Supervision of Professor Dr I.S. Zonneveld.

Methods and Study Area

This approach is based on the hypothesis that ibex range can be evaluated on the basis of snowfree winter habitat in the high mountains. Snowfree vegetated areas can be found on Landsat images which allows extensive areas to be searched easily. To test this hypothesis in the Gran Paradiso and Vanoise National Parks, predicted and actual ibex distributions were compared.

Landsat Imagery

Six Landsat scenes were obtained as computer compatible tapes, covering the mountain massif of the Vanoise-Gran Paradiso National Parks. The imagery comprised scenes of January, March, May, July and September (two years), and was obtained during mid-morning passes at approximately 09.30. Processing facilities provided by the ITC (International Institute for Aerial Survey and Earth Sciences, Netherlands) included their interactive operating system with a PDP 11/34, connected to a Ramtek colour monitor, a Gould electrostatic plotter and a Tektronix graphics terminal for displaying two-dimensional feature spaces.

False colour composites were generated for all six Landsat scenes, and visual interpretation of vegetation and terrain types was undertaken. An index of aerial green phytomass was calculated for three months (May, July and September). For this index a normalised Hadamard transformation was processed, which is considered to provide the most accurate indicator for a green leaf area index (Bronsveld & Luderus, 1982).

Preprocessing included radiometric and geometric corrections. Radiometric corrections adjusted brightness level of all pixels and eliminated seasonal variations, while geometric corrections enabled a superimposition of the three scenes. For the superimposition, a simple hybrid method was employed, using transparencies of Gould band 7 printouts, which by visual interpretation were registered to each other. A normalised Hadamard 1 transformation resulting in a green leaf area index for the three scenes was given the primary colours, blue, green and red.

Results

Ibex Wintering Areas

Ibex wintering areas were found on the Landsat scene of March 1976. False colour composites of this image show red-coloured patches of snow-free vegetation on sun-exposed slopes (Figure 24.2). In this way all 18 known major wintering locations in the

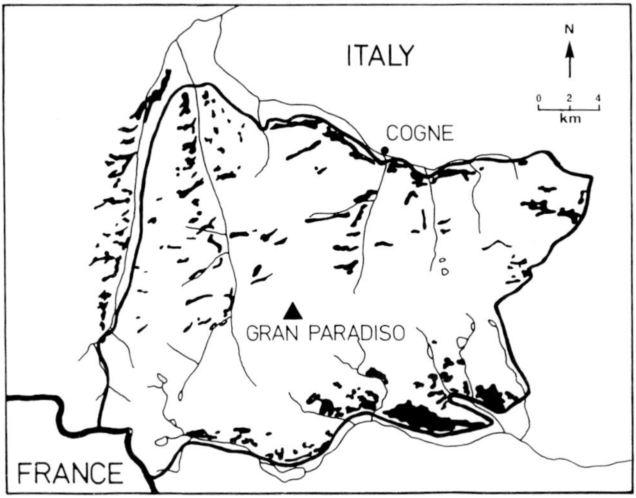

Figure 24.2: Predicted Winter Habitats of Ibex in Gran Paradiso National Park Through False Colour Landsat-Imagery. These winter habitats contain all the major ibex concentrations observed during winter

Gran Paradiso Park and an adjacent valley coincide with the reflectance of vegetation in band 7. Furthermore, all six known main wintering locations in the Vanoise National Park appeared to be overlapped by snow-free areas visible on the false colour composite of March. All known ibex winter locations coincide on this image with snow-free places, but of course not all snow-free places have ibex, but only those places where the animals are well protected and where the winter habitat is located on sun-exposed and generally steep slopes. The false colour composite for January showed, even in mid-winter, reflectance of green vegetation high above the tree line, indicating that in these places green forage is available the whole year round.

Green Biomass on a Multi-temporal Image

A multi-temporal image was created combining May, July and September. The sequential onset of vegetation growth can be seen, extending from lower to higher elevations and from sun-exposed to north-facing slopes. This image shows the extent of green vegetation cover in the mountain range at different times. It helps to explain seasonal displacements and the distribution patterns of ibex and indicated areas of possible conflict with animal husbandry (Wiersema, 1983).

Discussion

Ibex evolved in arid mountains with little snow. In the Alps this species was never widely distributed, as were chamois. In the Alps good ibex range is limited to mountains of little precipitation, where topography allows sufficient solar irradiation. It is in these patches of habitat that ibex can survive in winter. The method developed allows extensive mountain ranges to be searched for ibex winter habitat as a first step. A habitat suitability map should be the basis for further introductions. This method can ultimately be of use for finding valuable habitat for wildlife preserves in other high mountains of the world.

References

Bronsveld, M.C. & Luderus, F.J.D. (1982) *Analysis of Multi-temporal Data for the Identification of Land Use and Crops*, Technical report ITC, Enschede and IBM, Amsterdam-Madrid, vol. I, p. 38 and App. and Vol. II, p. 45

Wiersema, G.J. (1983) 'Analysis and Mapping of Ibex Habitat Using Digital Landsat Data', *ITC-Journal*, June

—— (In press) 'Project Ibex: Ibex Habitat Inventory and Mapping in the European Alps', *J. Mountain Research and Development*

PART FOUR

Management of Health and Disease

CHAPTER 25

Problems in the Management of Chamois in Captivity

H. WIESNER

Zoological Director, Münchner Tierpark Hellabrunn AG, Siebenbrunner Str. 6, D-8000 München 90 (FRG)

Abstract

Practical advice is given for the husbandry and management of chamois in captivity. The most common diseases and their treatment are discussed. The dosages of three drugs for a safe capture are reported.

Introduction

In Munich Zoo there is a long tradition, dating back in fact to its founding in 1928, of keeping chamois and other mountain ungulates, e.g. markhor, Nubian and alpine ibex, wild goats. Whereas the Caprinae sub-family is quite common in wild animal parks and zoos and their reproduction rate appears to be quite satisfactory everywhere, markhors and especially chamois are much more sensitive and difficult to manage. This may account for the fact that one rarely finds prosperous colonies of these species in captivity. At the time of writing, we have 11.11 markhors and 5.13 chamois in our collection, i.e. according to the International Zoo Yearbook, the biggest herds of either species in captivity the world over. The process of building up such complete colonies of high mountain ungulates was as stony and difficult as any one of these magnificent animals' biotopes. As the problems of nutrition, husbandry, management and diseases are, to a great extent, common to both species, I would like to summarise our experiences in this field and make some practical recommendations about the keeping of for example chamois.

General Management and Nutrition

One of the nastiest problems for mountain ungulates in captivity seems to be the overgrowing of hooves. Uncorrected hooves often result in malposition of joints and hypertension of tendons and ligaments which can then be irreversibly damaged. Using the old type of ordinary concrete enclosures, we had to trim hooves twice a year. Six years ago, however, we introduced a much more suitable surface for chamois, consisting of a thin layer of gravel with large areas of grass. As the gravel is sharp and rolls on the ground, the chamois' hooves are bound to get scraped and we have not had to trim them since then.

The surface area of our enclosure is ca. 4,240 m^2; the surrounding fence is 2 m high.

In captivity chamois can become very aggressive towards each other and behavioural interactions are very common. The dominant buck only tolerates young males until they reach the age of sexual maturation. Even between females, severe, aggressive interactions with regard to order of precedence can sometimes occur with fatal results. Therefore the structure of the surface of the compound should be formed in such a way that subordinate animals can easily hide from those holding a higher rank, when necessary. This can be achieved simply by adding rocks, a dry moat, trees or smaller hills. All the births in Hellabrunn Zoo occurred away from the visitors, behind a little range of hills, and during the first few days, the newborn kids could lie down with the protection of these hills. During the rutting season, the buck can become very harassing towards an individual female and chase her until she reaches the point of exhaustion. To avoid losses from this, we keep the female in a neighbouring enclosure for several days and test the buck's reaction daily. When the female seems to be fully oestrous, we bring her back to the herd and, in most cases, she is accepted again by the male. Normally mating occurs soon afterwards. On the other hand, when an individual female has once been attacked by several other females, reintroduction into the group has always been impossible and we have had to sell the animal. The reason for this behaviour between animals which have been together for several years, is unknown. In any case of handling, separating or driving the animals, we must always remember that chamois are very sensitive to stress and inclined to panic reactions, with a shock syndrome. Food should be available in several places to reduce competition at feeding and fresh portions given twice a day. Chamois are selective eaters and prefer branches, leaves and shoots. In the wintertime, we strongly recommend a daily provision of branches to provide the chamois with bark. Hay must be of first class quality and provided *ad libitum*. The average daily ration per animal has been set out in Table 25.1. Salt stones should always be available. The average body weight of our adults varies between 30 and 40 kg, indicating that they are in good condition and that their feeding is adequate.

Bacterial Infections

The greatest problem we had with our chamois was peracute septicaemia in kids during the first four weeks of life. The clinical symptoms were high temperature (41.5°C), convulsions and peracute death within a few hours. The mortality rate was very high: between 80 and 100 per cent. All the usual treatment proved useless. *Escherichia coli* strains 0.8, 086, 0117, could be isolated in all organs. Since 1975 we have vaccinated all our mountain ungulate kids with an oral vaccine (Baljer, 1975) produced from isolated bacteria strains taken from our dead young animals. This 'herd-specific' vaccine must be administered daily for the first 10 days of life. We built a special neighbouring enclosure for catching the young in order to avoid unnecessary stress during the capture. The effect of this vaccination has been excellent and has reduced the mortality in kids from colisepticaemia to zero.

In the years 1976/77, we suffered some losses of 6- and 13-year-old chamois due to acute bronchopneumonia caused by *Pasteurella multicida*; for several weeks previously, the animals had seemed to be in bad condition and had been separated from the herd. At the time we assumed that the nutrition was not suitable. Furthermore, this was probably the cause of various outbreaks of enterotoxaemia, and in a single case, of *Clostridium*

Table 25.1: Daily Food Ration for Chamois

Summer:	grass *ad libitum* 500 g pellets* 50 g mineral and trace element mixture 100 g concentrate** 500-1,000 g hay no fruit
Winter:	500-1,000 g corn silage 500 g hay 500 g brewers grains, dried 500 g pellets* 50 g mineral and trace element mixture 100 g concentrate**

pellets consisting of:		**concentrate consisting of*:	
25.00%	oat	20%	soy bean meal
21.00%	barley	30%	maize
16.00%	soy bean meal	40%	oat, crushed
10.00%	wheat bran	2%	brewers yeast
6.00%	alfalfa	2%	linseed expeller
6.00%	molasses	6%	dried sugar beet pulp
5.00%	wheat	100%	
4.00%	coconut cake		
2.00%	linseed cake		
1.50%	dicalcium phosphate		
1.00%	calcium carbonate		
0.55%	salt		
1.95%	vitamin-premis I.U.		
100.00%			

Vitamins:		
	47.000 I.E.	Vitamin A
	5.600 I.E.	Vitamin D_3
	312 mg	Vitamin E

septicum infection. We performed a vaccination twice a year by blowpipe with a polyvalent *Clostridium* vaccine (Covexin 8, Burroughs & Wellcome) over a period of two years. Since 1977 we have kept the animals in a larger enclosure, we have not repeated the vaccination and there have been no more instances of *Clostridium* infections. We think that the improvement in surroundings and more suitable nourishment has prevented any further outbreak of this polyfactorial disease.

Viral Infections

In 1977 we had an outbreak of an *Ecthyma contagiosa*-like syndrome after importing 0.2 animals from an animal park in Switzerland. Mostly kids of 2-3 months of age became heavily infected.

The first symptoms were only some grass blades sticking out at the lips. The lips were swollen with rhagades and bloody crusts around the mouth. A virus from the Papillon-Polyoma complex was suspected of being the cause but isolation of the virus was not possible. The lesions had been superinfected by *Staphylococcus aureus*. Treatment with antibiotics and daily washing with 0.2 per cent solution of hexachlorcyclohexan was effective.

Parasites

In all our mountain ungulates residual infestations of Strongylids like *Ostertagia*, *Trichostrongylus*, *Haemonchus*, *Chabertia*, *Oesophagostomum* and of *Capillaria*, *Trichuris* and lung worms (*Cystocaulus*, *Protostrongylus*, *Muellerius* and *Neostrongylus*) occur regularly. We control them by feeding 500 p.p.m. Mebendazole over a period of 2 weeks twice a year (Forstner, Kopp & Wiesner, 1977). In addition all our kids are dewormed by a regular injection of L-Levamisol 2-3 times during the first 8 months of life (Wiesner, 1975). With these precautions helminths are no longer a problem in keeping chamois.

On the other hand, coccidiosis is still a big problem with all high mountain ungulates (*Eimeria alpina*, *E. riedmülleri*, *E. rupicaprae*, *E. yakimoff-matschoulskyi*). Although these parasites are known as a specifically young animal disease in domestic stock, we even lost some chamois of more than 10 years of age from coccidiosis, after a stress situation. There is particular danger to kids from 8 weeks to 1 year old, as they become affected by severe diarrhoea. Severe electrolyte deficiency may provoke typical oedema in the throat region. Without treatment, the kids die within a few days. We have developed two therapeutic schedules for the treatment. In cases of an acute form of coccidiosis we give 2-5 g Socatyl-R (sulphathiazole plus formaldehyde) daily by throat tube for 3-5 consecutive days, with an additional dose of Theracanzan-R (sulphadimethoxin) (2-4 ml) per kid.

We did not get good results with the Amprolium 40 mg/kg for 14 days, sometimes recommended as a preventive measure. In our own field trials since 1975, we have found that 100 p.p.m. Monensin for 3 weeks is the best preventive treatment for coccidiosis in all mountain ungulates. By administering this treatment two or three times a year depending on the oocyst occurance in the faeces, we have had no more clinical symptoms of coccidiosis in our kids or losses from this disease. The excretion of oocysts decreases considerably, although some oocysts are always found in normally formed stools. Chamois do not accept a higher dose in their food. Monensin is known to be toxic in the equine family and too high a dosage has produced cardiomyopathy in deer (Amend *et al.*, 1980; Glover & Wobeser, 1983). With a dosage of 100 p.p.m. Monensin no side effects were observed in our chamois or in other mountain ungulates. This preventive treatment for coccidiosis and the oral vaccination against colisepticaemia seem to be the key to keeping a prosperous colony of chamois in captivity.

Immobilisation

Chamois are very sensitive to any kind of immobilisation and need on average much less drug dosage than other ungulates. Safe and efficient immobilisations in over 60 chamois were performed in Munich with the following combinations:

1. 'Hellabrunner mixture' (i.e. 125 mg Xylazin + 100 mg Ketamin per ml) (Wiesner, 1977): *subadult*: 0.04 ml in total; *adult*: 0.08 ml in total.
2. Immobilon® (*large animals*): 0.8 ml/adult animal (Wiesner, Rietschel & Gatesman, 1982)
3. 400 μg Carfentanyl + 2 mg Xylazin/adult animal (Wiesner *et al.*, in preparation).

References

Amend, J.F., Mallon, F.M., Wren, S.B. & Ramas, A.S. (1980) 'Equine Monensin Toxicosis: Some Experimental Clinicopathological Observations', *Compend. Continuing Educat.* II, *10*, 173-83

Baljer, G. (1975) 'Orale Vakzination neugeborener Kälber gegen *E. coli*-Erkrankungen', *Vortrag zum XX Welttierärztekongress*, Thessaloniki

Forstner, M.J., Kopp, H. & Wiesner, H. (1977) 'Weitere Untersuchungen uber die Entwurmung von Zoowiederkäuern mit Mebendazol', *Berl. Munch. Tierarztl. Wschr.*, *90*, 180-3

Glover, G.J. & Wobeser, G. (1983) 'Monensin Toxicity in a Captive White-tailed Deer (*Odocoileus virginianus*)', *J. Zoo. An. Med.*, *14*, 13-16

Wiesner, H. (1975) 'Zur Bestandsprophylaxe bei Hochgebirgstieren in Gefangenschaft', *Verhandl Ber. XVIII Intern. Symp. Erkrankungen Zootiere*, Innsbruck, 89-93

—— (1977) 'Zur Narkosepraxis mit dem "Blasrohrgewehr" ', *Kleintierpraxis*, *22*, 327-30

——, Rietschel, W. & Gatesman, T. (1982) 'Erfahrungen mit der Kombination von Immobilon® und Rompun® beim Zootier', *Zschr. Kölner Zoo Jahrgang*, pp. 47-55

CHAPTER 26

Parasitic Pneumonia in Chamois (*Rupicapra rupicapra* L.) in Slovenia

A. BIDOVEČ, S. VALENTINCIČ and MAJDA KUSEJ
Institute for Veterinary Hygiene and Pathology of the Game Biotechnical Faculty, University of Ljubljana, Ljubljana (Yugoslavia)

From 1974 to 1980 examinations of the parasitic fauna in 1,232 chamois lungs were carried out. Samples collected from practically all the regions occupied by chamois in Slovenia were included. The material was obtained from the annual culling, i.e. from 1 August to 31 December, and from elimination of surplus animals in the area of the Karavanke and Julijske Alps, which was carried out because of sarcoptic mange infestation of chamois.

The dissection of over 105 chamois found dead was also carried out. The lungs were examined by observation and palpation. These examinations revealed subpleural changes on both diaphragmatic parts of the lungs. These changes consisted of hard nodules a few millimetres wide, protruding above the surface of the organ, and the whole diaphragmatic apexes showed pathological changes. The lesions were never found diffused throughout the whole lobe of the lungs.

Samples were prepared by Skrjabin's method and, after several sedimentations, adult parasites were isolated, examined under Rechart's magnifying glass and microscopically determined.

Of the 1,232 lungs examined, 1,023 (83.03 per cent) were infected by the above mentioned parasites. The parasites and their frequency expressed as a percentage of the total number of infected lungs are shown in Table 26.1.

The invasion intensity was estimated quantitatively as: (1)

Table 26.1: Distribution of Lung Parasites in the Examined Lungs

Species of lung parasites	Number of infected lungs	%
Muellerius capillaris	276	26.98
Neostrongylus linearis	199	19.45
Cystocaulus ocreatus	127	12.42
Protostrongylus raillieti	104	10.17
Dictyocaulus filaria	5	0.49
Protostrongylus rufescens	3	0.29
Protostrongylina larva	309	30.20
Total	1,023	100.00

single parasitic nodules; (2) extensive nodule invasion in the lungs; and (3) severe parasitic pneumonia.

The data on the age of the animals whose lungs were examined enabled us to correlate the intensity of invasion to the age of the animals (Figure 26.1).

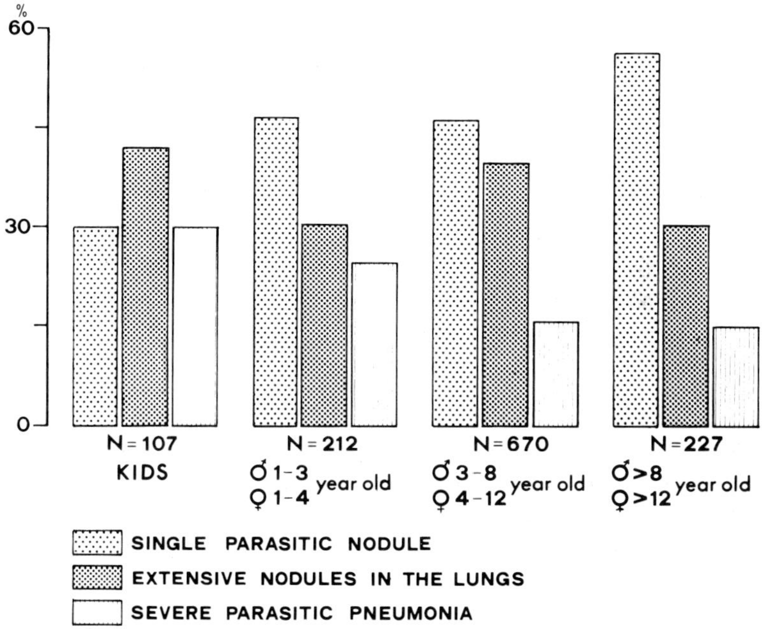

Figure 26.1: Invasion Intensity in Relation to Sex and Age

From the epizootological point of view, the spreading of strongyloidosis among chamois in Slovenia is aetiologically non-specific and can be found in domestic sheep and goats as well as in introduced ibex (*Capra ibex*) and moufflon (*Ovis musimon*). On the other hand, the extension and intensity of invasion in these animals only cause a relatively low mortality.

The occurrence of an intense parasitic pneumonia as well as other parasitoses and some viral infections may be related to a biological imbalance between the population size of wildlife species and the carrying capacity of their habitat. This is especially true in regions within the low forest zone, where the ecological conditions for the development of infective larvae are more favourable than those in the alpine ranges where chamois normally live.

Tourism and other man-induced disturbances may also be responsible for the decrease in size and quality of the chamois population, as they alter the animals' living area.

CHAPTER 27

The Impact of Mange on Chamois in Bavaria

CHRISTINE MILLER
Institute of Wildlife Research, University of Munich, Amalienstrasse 52, D-8000 München 40 (FRG)

Scabies of chamois is caused by the sarcoptic mite, *Sarcoptes rupicaprae*. From the Northern Alps, between the Austrian border to the East and the Inn river to the West, and south-east Bavaria, the mites are now on the verge of expanding to South Tyrol, north-east Italy and Yugoslavia (Figure 27.1).

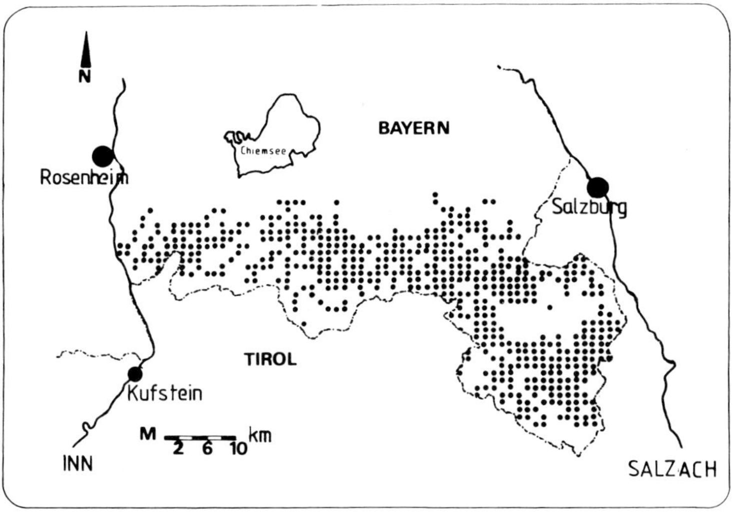

Figure 27.1: Occurrence of Chamois Scabies in the East Alps

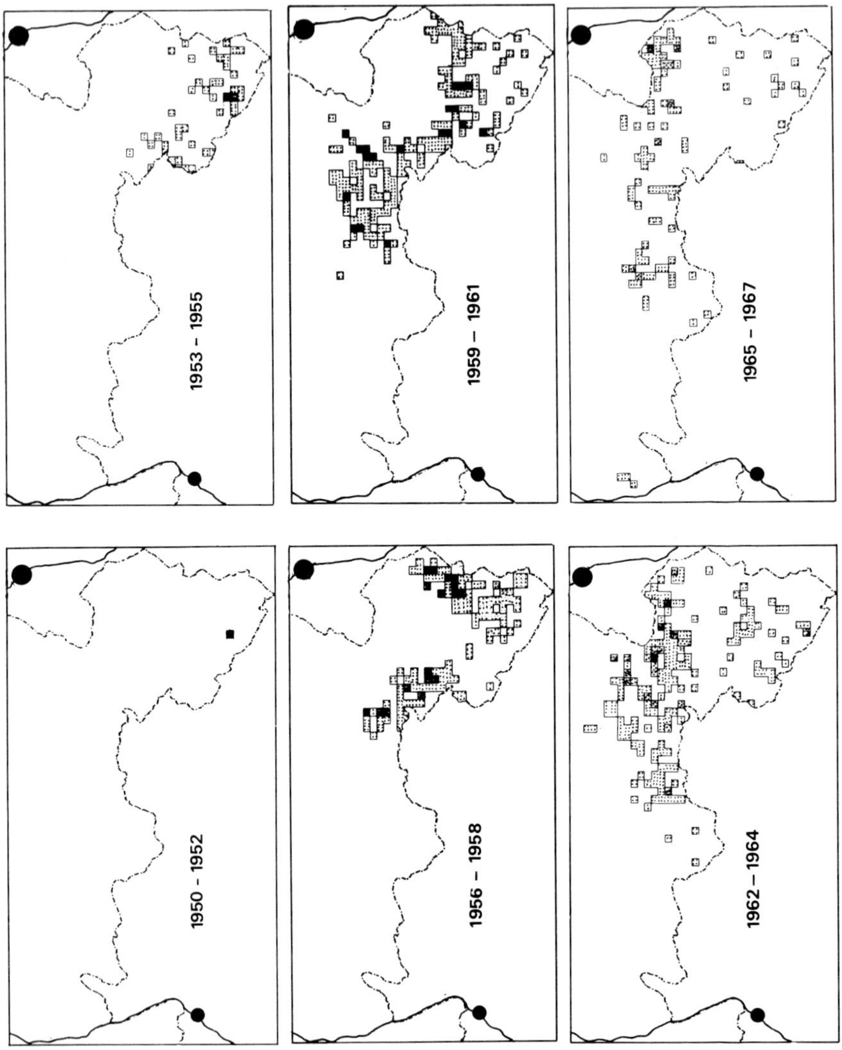
1953 - 1955
1959 - 1961
1965 - 1967
1950 - 1952
1956 - 1958
1962 – 1964

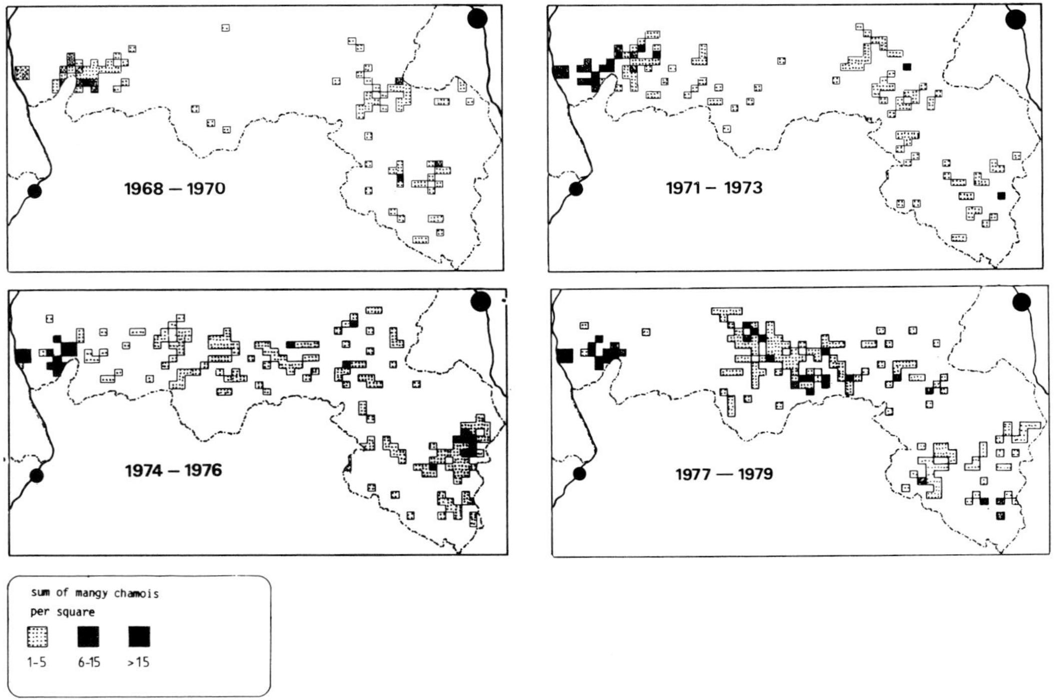

Figure 27.2: The Spreading of Scabies in Bavaria from 1950 to 1979. Each map sums up the casualties over three years

The dynamics of the colonising mites were studied on the Bavarian chamois, which were again stricken by scabies 30 years ago, after a latency of more than 100 years.

After the first occurrence of scabies in Bavaria in 1949, the local Hunting Managers recorded the site, age and sex of 3,205 chamois, which died of this epidemic. The discovery sites were plotted in quadrats of one square kilometre, over all the diseased area.

To document the expansion, a map was drawn for every three year period. The number of chamois found dead was recorded in the corresponding quadrats of the maps. Then, the data were analysed by computer. Figure 27.2 shows the advance of the epizootics. Scabies first appeared in the south of the area, later on spreading out to the west. The westernmost regions were reached in the late 1960s. The Inn river has stopped a further westward expansion of the epidemics up to now.

Three management districts were stricken in succession. Their respective annual scabies casualties are compared in Figure 27.3. The first severe outbreak is followed by a relatively rapid regression of the disease. The number of scabby chamois then remains low for several years; afterwards, a new outbreak may occur. Figure 27.4 shows the distribution of dead animals — except kids — by sex and age. Half of the animals die before they are 5 years old; most of them when they are 3 and 4 years old.

A typical chronological pattern of colonisation stands out: a first wave overruns the chamois population, which collapses. The mites persist in small numbers in the reduced chamois stock, but the mortality rate may still keep fairly high occasionally. The subsequent epidemic outbreaks are however smaller than the first one, but there were apparently two outbreaks in region no. 3 (Figure 27.3).

The age distribution of the scabby chamois includes a majority of 3- to 5-year-old individuals. These age classes are probably under a particularly severe stress because of their first pregnancy and lactation, or rut. Thus, their susceptibility to scabies is increased. The old animals dying of scabies are mainly females. The population, however, probably does include more females, since there is a greater hunting pressure on fully adult males. So the scabies deaths in these cases may be comparable to the age structure of the real population.

From 1968 to 1981, when fewer animals died of scabies, the

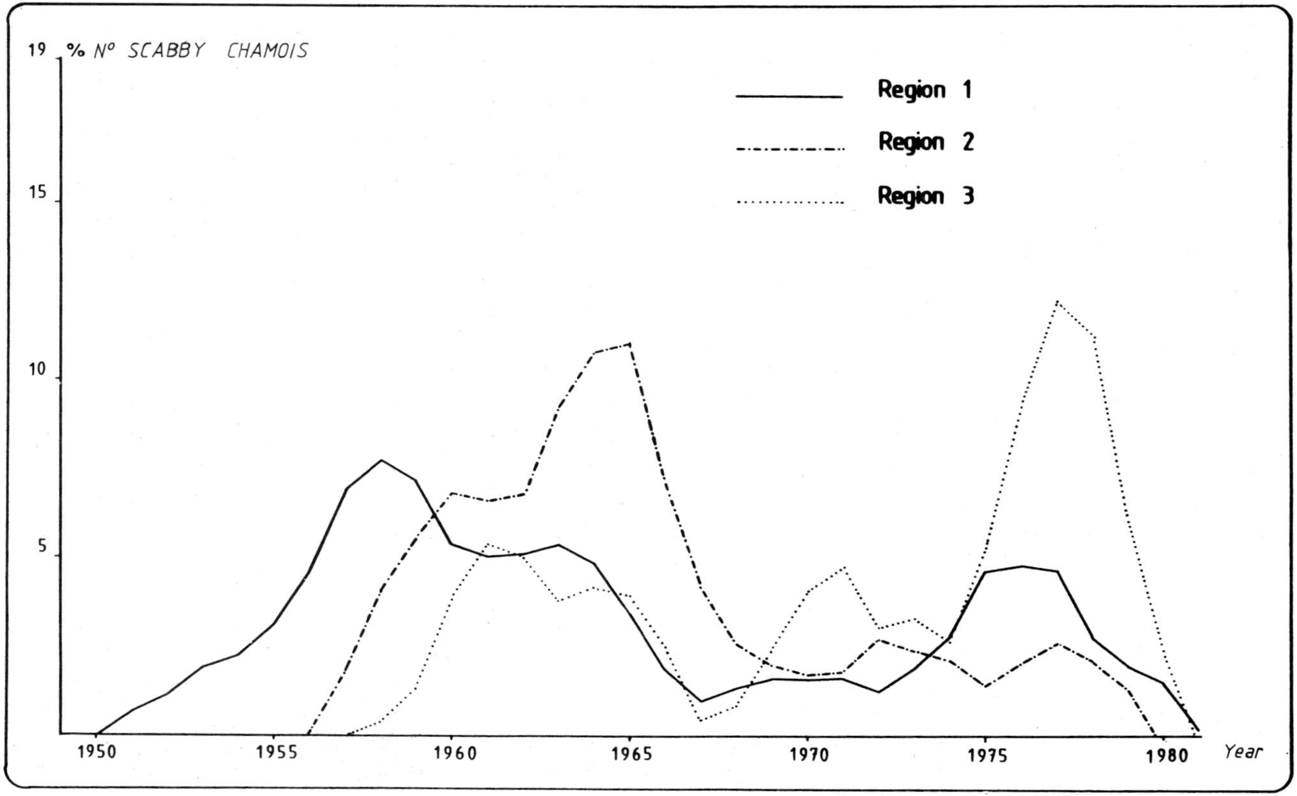

Figure 27.3: Outbreaks and Regression of Scabies in Three Regions since 1950, Expressed by the Annual Percentage of Sick Animals

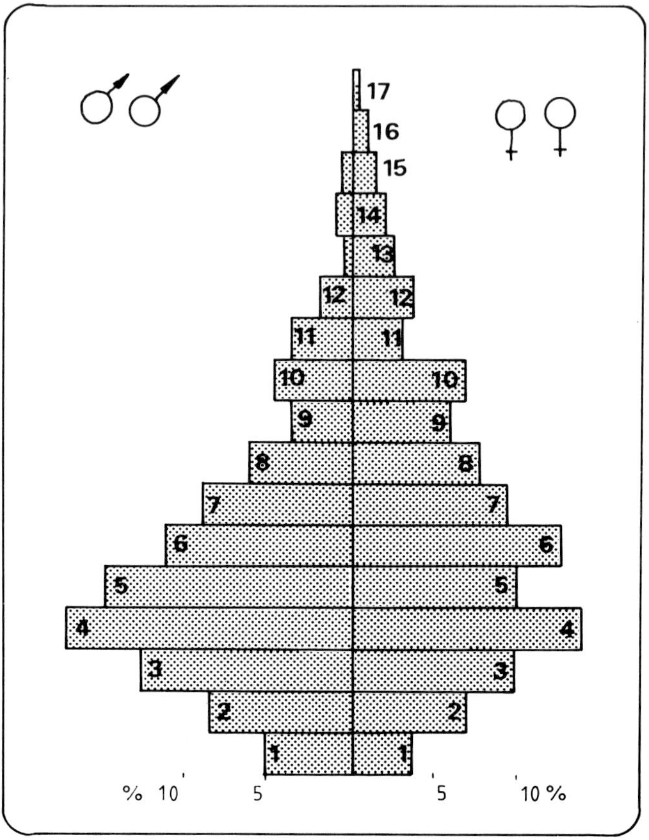

Figure 27.4: Age Distribution of Females and Males During the First Infection Wave from 1950 to 1966 (*left diagram*) and the Regression Period from 1968 to 1981 (*right diagram*)

age frequencies proved somewhat different. Stochastic influences on the transmission of mites and/or the sickness of an individual chamois may have become more significant.

A new system of recording the disease is being worked out at the Institute of Wildlife Research in Munich, which will allow the detection of the actual phase of the parasitosis and the prediction of the next ones.

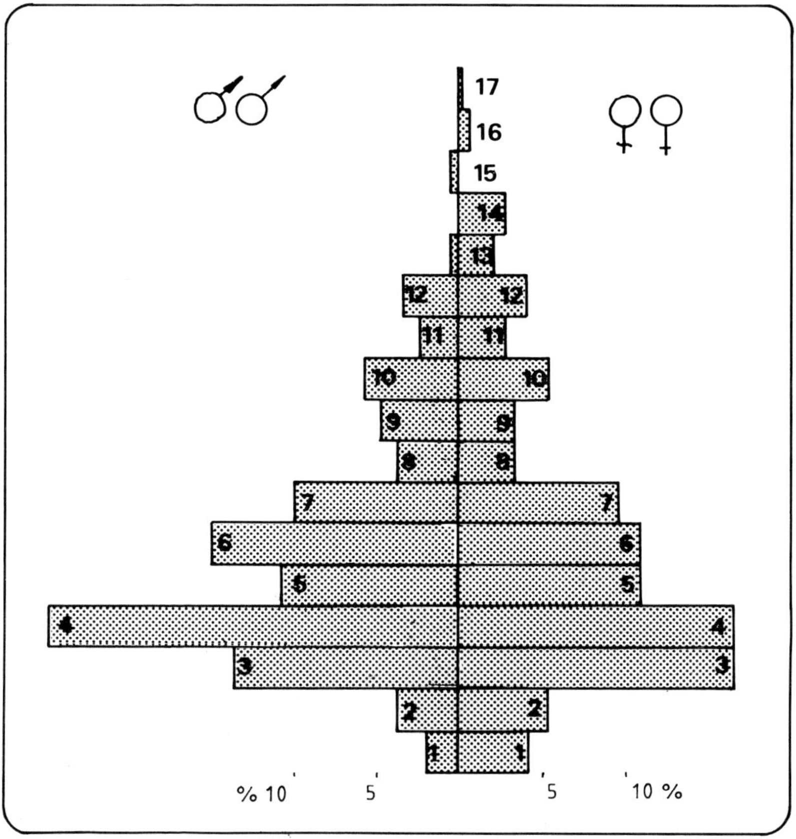
17
16
15
14
13
12
11
10
9
8
7
6
5
4
3
2
1
% 10
5
5
10 %

CHAPTER 28

Investigations on the Metabolism of Chamois Suffering from Sarcoptic Mange

FRIEDA TATARUCH, THEODORA STEINECK and K. ONDERSCHEKA

Forschungsinstitut für Wildtierkunde der Veterinärmedizinischen, Universität Wien, Wien (Austria)

Abstract

A total of 133 chamois were shot in different biotopes in Austria and Slovenia. Using biochemical, histological and botanical analyses, differences between the metabolic state of healthy chamois and chamois infested with sarcoptic mange were determined. It could be shown that sick animals were in a catabolic phase, its degree depending on the stage of the disease. Remarkable differences were found in vitamin A, zinc and iron metabolism, in the activity of the adrenal cortex and in feeding selection as well as in other factors.

Introduction

This paper reports on some results obtained in research on sarcoptic mange of chamois. This project was performed in collaboration with the Institut za Zoohigieno za Patologijo Divjih Zivali, University of Ljubljana. Our purpose was to check the metabolic situation of mangy chamois in relation to healthy individuals. By analysing samples of animals from different biotopes we tried to find out if there were biotope-dependent factors, favouring the infestation of sarcoptic mange.

Material and Methods

Of 133 chamois shot in different biotopes in Austria and Slovenia, there were three groups of animals: (1) chamois suffering from mange (N=33); (2) healthy chamois living in the same area (N=45); (3) healthy chamois from biotopes where no outbreak of mange had been recorded in recent years (N=55).

We carried out biochemical analyses of the rumen contents, liver, blood serum, urine, hair, ribs and faeces. Adrenals and thyroids were examined histologically. The botanical composition of the rumen contents and the burden of the chamois by environmental contaminants were also determined.

Results and Discussion

Liver and blood analyses proved that chamois suffering from mange are in worse condition than healthy animals. At the beginning of the disease the catabolic situation of the organism is characterised by a decline in the concentration of glycogen and the body pool of carbohydrates, whereas the amount of fat in the liver does not change (Table 28.1). The values of urea, glucose and total lipids in the blood also remain unchanged. In the next stage the body mobilises fat reserves from the peripheral parts into the liver. This results in a rise in the concentration of fat in the liver, glycogen remaining at a low level. The last stage of the disease is, finally, characterised by low concentrations of glycogen and fat in the liver accompanied by a remarkable decrease of total lipids in the blood (Table 28.1).

The high value of urea in the blood is an indication of the utilisation of the body's last energy deposits, the protein storage, with nitrogen in the urine showing a parallel course (Table 28.3). The high grade catabolism is also shown by the low concentration of glucose in blood, indicating a very reduced locomotory activity of the sick animal. More evidence of the malnutrition of infested chamois can be found in the significantly lower concentration of calcium (Spörri & Stünzi, 1969) in serum, in comparison with healthy animals.

Also food selection is remarkably influenced by the reduced ranging activity of the sick chamois: the botanical composition of the rumen contents shows a significantly higher proportion of shrublets and conifer needles, whereas the amount of grasses is only two-thirds that found in the rumen contents of healthy chamois (Table 28.2). The concentration of crude protein is also lower in sick animals (Table 28.3).

Table 28.1: Changes of Some Physiological Parameters During Different Stages of Mange

	Glycogen/Liver (mg/g d.w.)			Fat/Liver (% d.w.)			Urea/Serum (mg/100 ml			Total Lipids/Serum (mg/100 ml)		
	n	x̄	± s	n	x̄	± s	n	x̄	± s	n	x̄	± s
No signs of infection	31	43.5	29.8	31	9.28	4.18	21	23.6	13.6	19	421	136
Early stage	9	29.1	17.4	9	8.53	0.63	5	29.2	20.8	5	365	162
Advanced stage	6	7.5	12.2	5	12.24	2.1	3	41.1	15.4	3	442	185
Very advanced stage	12	1.1	3.0	11	6.98	1.82	10	80.3	32.0	10	200	102

Table 28.2: Botanical Composition of Rumen Contents (%)

		Group 1 (n=27)	s.l.	Group 2 (n=26)	s.l.	Group 3 (n=23)
Grasses	x̄	38.7	0.0	61.5	77.9	63.3
	± s	19.4	(0.0)	23.3		20.6
Leaves	x̄	4.7	84.5	5.1	2.0	0.9
	± s	9.3	(4.7)	6.2		1.5
Needles	x̄	21.8	0.1	7.0	77.7	6.2
	± s	17.7	(0.0)	11.2		7.9
Shrublets	x̄	19.4	7.1	11.8	25.1	17.5
	± s	17.3	(70.7)	12.2		19.9
Herbs	x̄	12.1	71.3	10.4	6.5	4.3
	± s	16.6	(3.1)	15.1		6.3
Mosses	x̄	0.2	72.7	0.3	7.1	0.1
	± s	0.7	(19.6)	0.6		0.2
Unidentified	x̄	3.0	56.2	3.6	77.2	3.1
fibres	± s	3.3	(94.8)	4.4		7.3

Table 28.3: Physiological Parameters of Mangy Chamois in Relation to Healthy Animals

		Group 1	s.l.	Group 2	s.l.	Group 3
Rumen cont.						
Crude prot.	n	25		29		22
(% d.w.)	x̄	15.78	18.7	17.11	50.8	16.55
	± s	4.30	50.5	2.59		3.39
Iron	n	23		29		21
(mg/kg d.w.)	x̄	256.3	31.7	218.4	98.7	217.8
	± s	157.5	42.4	94.6		158.8
Liver						
Glycogen	n	24		31		22
(mg/g d.w.)	x̄	10.0	0.0	43.5	16.5	33.0
	± s	14.8	0.0	29.8		21.6
Fat	n	22		31		21
(% d.w.)	x̄	8.64	49.5	9.28	46.0	8.59
	± s	2.63	95.3	4.18		2.48
Nitrogen	n	24		31		22
(% d.w.)	x̄	12.42	3.6	11.92	67.2	11.99
	± s	0.95	5.6	0.77		0.45
Vitamin A	n	23		28		21
(i.E./g d.w.)	x̄	2,289	0.9	1,648	11.5	1,999
	± s	974	32.3	528		879
Zinc	n	23		31		21
(mg/100 g ash)	x̄	448.8	2.5	255.4	37.6	270.0
	± s	385.3	3.9	52.3		65.5
Iron	n	23		31		21
(mg/100 g ash)	x̄	1,195	1.9	761.5	71.9	804.4
	± s	751	3.3	455.7		360.2
Urine						
Nitrogen	n	7		10		5
(mg/100 ml)	x̄	885.1	3.0	595.2	19.8	804.0
	± s	248.2	64.7	244.5		350.0
Ribs						
Iron	n	16		23		17
(mg/100 g ash)	x̄	17.84	4.0	12.70	2.6	16.55
	± s	8.56	61.4	4.66		5.85

Table 28.3 (cont.)

		Group 1	s.l.	Group 2	s.l.	Group 3
Faeces						
Iron	n	25		23		15
(mg/kg d.w.)	x̄	463.9	30.5	380.5	12.4	292.4
	± s	358.5	4.7	171.6		163.3
Blood-serum						
Calcium	n	18		21		19
(mg/100 ml)	x̄	10.06	1.3	11.57	29.7	10.98
	± s	1.55	7.6	1.99		1.50
Copper	n	16		19		17
(μg/100 ml)	x̄	80.0	0.4	56.8	46.4	50.6
	± s	23.7	0.4	20.3		29.9
Urea	n	17		21		20
(mg/100 ml)	x̄	66.54	0.0	23.61	53.2	26.6
	± s	38.2	0.1	13.6		16.8
Glucose	n	8		9		13
(mg/100 ml)	x̄	87.4	7.9	220.7	47.4	164.4
	± s	78.2	17.5	189.4		169.6
Total lipids	n	17		21		20
(mg/100 ml)	x̄	262.0	0.2	421.0	12.0	506.0
	± s	141.0	0.0	136.0		185
Vitamin A	n	9		10		10
(μg/100 ml)	x̄	14.5	2.3	27.9	8.2	18.1
	± s	5.3	15.8	15.3		5.3

abbreviations:
Group 1 = mangy chamois
Group 2 = healthy chamois in a biotope with mange
Group 3 = healthy chamois in a biotope without mange
n = number of samples analysed
x̄ = arithmetic mean
± = standard deviation
s.l. = significance level
Values in parentheses are valid for comparison between group 1 and group 3
d.w. = dry weight

The impact of mange on the chamois organism is also expressed by remarkable changes in the adrenal cortex, leading to an enlargement of the *zona fasciculata*, to the disadvantage of the *zona glomerulosa* and *reticularis*.

Another significant difference between sick and healthy

chamois was found in the distribution of vitamin A in blood and in the liver, which is the main storage organ for this substance (Table 28.3). Vitamin A is essential for the maintenance of normal epithelia including epidermis. Sick chamois serum contained significantly less vitamin A than healthy animals in mangefree biotopes.

The serum of non-infested chamois in the biotopes with occurrence of mange had even higher levels of vitamin A than other healthy animals (Table 28.3). This might be one of the reasons why some chamois did not show any symptoms of mange, although we could find mites on their skin.

In contrast to the results obtained from serum analyses, the concentration of vitamin A in the liver of sick chamois was significantly higher than in the liver of healthy ones. The livers of the sick animals also contained significantly higher amounts of zinc, which plays an important role in the transport of vitamin A. Therefore, it can be assumed that the mobilisation of vitamin A from its storage organ, the liver, into the blood and, in consequence, to the skin, is reduced or hindered in sick animals.

Furthermore, sick animals showed significantly higher concentrations of iron in the liver than healthy animals. Also ribs, rumen contents and faeces of animals suffering from mange contained significantly more iron. This might indicate that there is a better supply of this element. Yet, this contradicts the haematological results, which show a lower value for erythrocytes and haemoglobin in the blood of sick chamois.

The concentration of copper in the serum of these animals was significantly higher. This is usually explained as the organism response to an infection (Buddecke, 1978). On the other hand, pigs reared on an iron-deprived diet showed less severe lesions when infected experimentally with sarcoptic mites than pigs with a normal supply of iron (Sheahan, 1975). The question as to whether a higher iron-intake favours the outbreak of mange in chamois, too, will be the subject of further investigations.

References

Buddecke, E. (1978) *Pathobiochemie*. De Gruyter, Berlin & New York

Sheahan, B.J. (1975) 'Pathology of Sarcoptes Scabies Infection in Pigs', *J. Comp. Pathol.*, *85*, 87-95

Spörri, H. & Stünzi, H. (1969) *Pathophysiologie der Haustiere*. P. Parey, Berlin & Hamburg

CHAPTER 29

Occurrence of Pulmonary and Gastrointestinal Nematodes in the Abruzzo Chamois

GABRIELLA CANCRINI,[1] ALBERTINA IORI,[1] L. ROSSI[2] and R. FICO[3]

1. Istituto di Parassitologia, Università di Roma, Roma (Italy)
2. Cattedra di Malattie Parassitarie, Facoltà di Medicina Veterinaria, Torino (Italy)
3. Parco Nazionale d'Abruzzo, CSEA, 67032 Pescasseroli (AQ) (Italy)

In the National Park of Abruzzo there are about 400 Apennine chamois. This remnant population is considered very interesting from many points of view (genetical, anatomical, ethological, etc.) because of its long geographic isolation (e.g. Lovari, 1984).

In this note we outline the preliminary results of research carried out for the purpose of describing the parasitological condition among these animals.

Between 1976 and 1981 the organs of the respiratory and digestive systems of 21 chamois (age ranging between 3 and 10 years) were examined at the Institute of Parasitology of Rome.

The sample could not be larger as no kind of harvesting is allowed in the study area. Moreover, collection of parasites proved quite difficult as some of the animals were found only several days after their death. Consequently a complete examination of the respiratory and digestive systems was only possible in the case of six chamois; for the other 15 animals, single organs were examined (6 abomasa, 5 small intestines, 1 large intestine and 7 lungs).

The dissection of the organs and the collection of the parasites were carried out so that all nematodes could be picked up.

Respiratory System

The presence of male specimens of *Protostrongylus rupicaprae* (49 males in all) was demonstrated in seven chamois; in three other animals we collected female nematodes of the genus *Protostrongylus* whose identification could not be completed.

Digestive System

The following species of nematodes were identified: *Marshallagia marshalli*, *Nematodirus abnormalis*, *Oesophagostomum venulosum*, *Ostertagia circumcincta*, *Ostertagia lyrata*, *Ostertagia occidentalis*, *Ostertagia trifurcata*, *Teladorsagia davtiani*, *Trichostrongylus axei*, *Trichostrongylus capricola*, *Trichostrongylus colubriformis*.

The parasitic burden was always quite low (680 males in all). Finally, protozoa belonging to the genus *Sarcocystis* were found in the hearts of two chamois out of eight.

Both the number of nematode species identified and their density are surprisingly low when compared with other chamois populations (e.g. Dollinger, 1974; Pérdrix *et al.*, 1976; Polley & Hörning, 1977; Balbo *et al.*, 1978).

Nevertheless, the exiguity of the sample and the lack of findings from the younger age classes (kids and yearlings) prevent us from drawing definitive conclusions about the parasitological condition of the Abruzzo chamois.

References

Balbo, T., Costantini, R., Lanfranchi, P. & Gallo, M.G. (1978) 'Raffronto Comparativo della Diffusione dei Nematodi Gastro-intestinali nei Ruminanti Domestici (*Ovis aries* e *Capra hircus*) e nei Ruminanti Selvatici (*Capra ibex* e *Rupicapra rupicapra*) delle Alpi Occidentali', *Parassitologia*, *20*, 131-7

Dollinger, P. (1974) 'Beitrag zur Kenntnis des Endoparasitenspektrums des Gamswildes in der Schweiz', *Z. Jagdwiss.*, *20*, 115-18

Lovari, S. (1984) *Il Popolo delle Rocce*, Rizzoli, Milan

Perdrix, J., Sarrazin, C., Reydellet, M., Gindre, R. & Bastide, J. (1976) 'Observations Préliminaires sur L'état Sanitaire du Chamois et du Mouflon dans les Hautes-Alpes', *Bull. Soc. Sci. Vet. Med. Comp. Lyon.*, *78*, 89-92

Polley, L. & Hörning, B. (1977) 'The lungworm *Spiculocaulus austriacus* (Gebauer, 1932) Dougherty and Goble, 1946 in chamois (*Rupicapra rupicapra*) in Switzerland', *Rev. Suisse Zool.*, *84*, 675-80

CHAPTER 30

Parasitological and Pathological Observations on the Chamois in the Western Alps

T. BALBO, P. LANFRANCHI and L. ROSSI
Cattedra di Malattie Parassitarie, Facoltà di Medicina Veterinaria, Torino (Italy)

Abstract

The aim of this paper is to provide a review of the main pathogens found in chamois of the Italian Western Alps during the last 15 years. Our findings included: (A) the possibility of transmission of both pulmonary and gastrointestinal nematodes between different species of wild ruminants (chamois and ibex) as well as between wild and domestic ones (chamois, sheep and goats); (B) the presence of a 'mountain ungulate — fox' cycle for at least one species of protozoa (Sarcocystidae); (C) a significant correlation between the population density of chamois and the mortality rate derived from some diseases, e.g. infectious kerato-conjunctivitis.

Introduction

According to the most recent estimates (Perco & Tosi, in press) there are 22,000 chamois living in the Western Alps. These ungulates are not found evenly distributed along the entire mountain range but are often concentrated within protected areas and private hunting preserves. The sometimes high density of these chamois (up to as many as 25 animals/100 ha) is in itself an initial worry from a sanitary point of view. But these problems become more serious if we consider that the same areas are common to other wild ruminants (ibex, red deer, roe deer, moufflon) and in the summer time are also heavily grazed by domestic ruminants.

As a consequence dangerous situations of promiscuity may arise, which can, under certain conditions, have unpredictable repercussions for both the wild and domestic animals.

In this lies the interest of studying the pathogens present in these biotopes, their diffusion and the possibility of transmission from one species to another.

Materials and Methods

From 1966 onwards parasitological surveys of chamois, sheep and goats from the whole Western Alps and ibex from the National Park of Gran Paradiso have been carried out. In total 220 chamois, 59 ibex, 88 sheep and 12 goats were examined. For the techniques of collection and identification of the parasites we refer to previous papers (Balbo, Costantini & Peracino, 1973, 1975).

Microbiological, anatomical and histopathological investigations of the eyes of 29 chamois and one ibex were carried out during a recent outbreak of infectious keratoconjunctivitis. Information regarding other diseases came from our routine diagnostic activities and also from original papers by researchers in the Faculty of Veterinary Medicine in Turin.

Results and Discussion

In the study area the respiratory system of chamois can host the following species of nematodes (Balbo *et al.*, 1975; Balbo & Costantini, 1976; Biocca, Balbo & Constantini, 1976): *Spiculocaulus austriacus*, *Protostrongylus rufescens*, *Protostrongylus hobmaieri*, *Protostrongylus rupicaprae*, *Protostrongylus rufescens* var. *boevi*, *Dictyocaulus filaria*.

The first three species are common to the ibex. As a rule the percentage of chamois infected, the variety of parasite species and the parasitic burden are greater in the more densely populated areas.

In the digestive system the following species of nematodes can be found (Balbo *et al.*, 1973, 1978; Biocca *et al.* 1974, in press): *Chabertia ovina*, *Haemonchus contortus*, *Marshallagia marshalli*, *Nematodirus abnormalis*, *Nematodirus davtiani alpinus*, *Nemat-*

odirus filicollis, *Nematodirus helvetianus*, *Nematodirus ibicis*, *Nematodirus oiratianus*, *Nematodirus rupicaprae*, *Nematodirus spathiger*, *Oesophagostomum venulosum*, *Ostertagia circumcincta*, *Ostertagia lyrata*, *Ostertagia occidentalis*, *Ostertagia ostertagi*, *Ostertagia trifurcata*, *Skrjabinema ovis*, *Skrjabinema rupicaprae*, *Trichostrongylus axei*, *Trichostrongylus colubriformis*, *Trichostrongylus vitrinus*, *Trichuris globulosa*, *Trichuris ovis*, *Trichuris skrjabini*. Of these species 18 are also present in the ibex (Table 30.1)

Table 30.1: Nematodes Found in the Digestive System of Sheep, Goats, Ibex and Chamois

	Frequency of occurrence in host %			
	Ovis aries	*Capra hircus*	*Capra ibex*	*Rupicapra rupicapra*
Bunostomum trigonocephalum	34	33	—	—
Chabertia ovina	30	25	50	30
Cooperia turticci	1	—	—	—
Haemonchus contortus	34	42	3	29
Marshallagia marshalli	7	—	75	82
Nematodirus abnormalis	2	—	—	1
Nematodirus davtiani alpinus	—	—	68	23
Nematodirus filicollis	46	33	53	60
Nematodirus helvetianus	5	8	—	4
Nematodirus spathiger	16	—	11	7
Oesophagostomum venulosum	2	8	—	4
Ostertagia circumcincta	50	91	32	49
Ostertagia (G.) *lyrata*	1	—	28	4
Ostertagia (G.) *occidentalis*	—	—	64	64
Ostertagia ostertagi	6	8	—	7
Ostertagia trifurcata	26	58	7	4
Skrjabinema ovis	3	—	3	1
Skrjabinema rupicaprae	—	—	3	3
Trichostrongylus axei	19	33	14	15
Trichostrongylus colubriformis	35	75	21	9
Trichostrongylus vitrinus	21	33	39	7
Trichuris globulosa	—	—	—	3
Trichuris ovis	3	—	11	15
Trichuris skrjabini	1	—	14	11

As with the pulmonary nematodes, the degree of infection is higher in animals from more densely populated areas. Many species of these nematodes also infect the sheep and goats which graze on the Western Alps; at least 20 species have been found in these sheep and 12 in the goats (Table 30.1). From this it would seem quite clear that mountain ungulates can be a reservoir of parasites for domestic ruminants and vice versa. This possibility has been confirmed by the experimental infection of sheep and goats with nematode larvae from faecal cultures taken from chamois and ibex. The chamois of the National Park of Gran Paradiso (PNGP) are frequently seen to be infected with protozoa of the family *Sarcocystidae*, for which the red fox has been shown to be the main host (Biocca *et al.*, 1975). Sarcocysts have been discovered in more than 80 per cent of the heart sections examined, both in the myocardial and in the conduction tissue (Cornaglia *et al.*, 1980). Sometimes insects of the genus *Melophagus* and more rarely some light infestation of ticks are to be seen.

Of particular interest is the absence, at the moment, of sarcoptic mange.

Papillomatosis and 'chamois blindness' are the main infectious diseases that it has been possible to observe. Papillomatosis occurs sporadically or enzootically in various regions of the Western Alps, where it mainly hits chamois less than one year old in the late winter and spring. In this context Maglione, Ginanni & Sacco (1969) demonstrated the biological identity of the disease with the 'contagious ecthyma' of sheep and goats.

Much more important is the outbreak of infectious kerato-conjunctivitis which exploded in the late summer of 1981 and still continues. This disease spread rapidly over the slopes of Monte Rosa westward across the PNGP as far as the National Park of Vanoise in France.

Until May 1983, on the Italian side, more than 500 sick chamois and 20 ibex have been shot or found dead, with a distinct prevalence of females even when at the outset the sex-ratio seemed balanced (Table 30.2). The mortality rate increased with the density of the population (Table 30.3).

Doubts still remain as to the causal agent of the infectious keratoconjunctivitis. In scrapings from the conjunctival epithelium we could demonstrate 'inclusion bodies' the morphology of which points towards the probable intervention of a *Chlamydia*; our results are confirmed in the research of Spanish and French authors who have studied the same problem in the isard (Belda & Ferrando, in press; Tournut, in press). Certainly a

Table 30.2: Infectious Keratoconjunctivitis: Mortality Rate in Relation to Sex and Sex-ratio

	Males (%)	Females (%)	Sex-ratio (males/females)
Grant Avert	12.6	87.4	1:2.6
Dondenaz	14.9	85.1	1:3
PNGP	6.9	93.1	1:1.2

PNGP = Gran Paradiso National Park.

Table 30.3: Infectious Keratoconjunctivitis: Mortality Rate and Density

	Chamois/100 ha.	Mortality (%)
Grant Avert	21.0	25.2
PNGP (Cogne)	12.5	7.3
Dondenaz	10.1	4.1
PNGP (Orco-Soana)	4.7	1.9

PNGP = Gran Paradiso National Park.

Streptococcus-β-haemolytic of the C group seems to play an important part in the evolution of the eye damage; in fact we isolated this pathogen in roughly 80 per cent of the diseased chamois' eyes whereas it is absent from the eyes of healthy chamois shot in areas free from infectious keratoconjunctivitis. By subconjunctival inoculation of this microrganism it has been possible to reproduce a form, although light, of keratoconjunctivitis in a kid.

In the chamois of the Western Alps a certain number of deaths are due to a necrotic bronchopneumonia which attacks animals in poor condition in winter time (Valenza *et al.*, 1976). The aetiology of this disease remains unknown.

Finally organic diseases such as nephropathies, hepatopathies, myocardiopathies and arteriosclerotic lesions have been repeatedly described in the chamois of the PNGP (Guarda & Calliera, 1961: Guarda, Guarda & Biolatti, 1980; Guarda, Guarda & Cornaglia, 1980).

References

Balbo, T. & Costantini, R. (1976) 'Sulla Disuguaglianza degli Spicoli nel Genere *Spiculocaulus*', *Parassitologia*, *18*, 67-70

Balbo, T. Costantini, R., Lanfranchi, P. & Gallo, M.G. (1978) 'Raffronto Comparativo della Diffusione dei Nematodi Gastro-intestinali nei Ruminanti Domestici (*Ovis aries* e *Capra hircus*) e nei Ruminanti Selvatici (*Capra ibex* e *Rupicapra rupicapra*) delle Alpi Occidentali', *Parassitologia*, *20*, 131-7

Balbo, T., Costantini, R. & Peracino, V. (1973) 'Indagine sulla Diffusione dei Nematodi Gastro-intestinali nello Stambecco (*Capra ibex*) e nel Camoscio (*Rupicapra rupicapra*) nel Parco Nazionale del Gran Paradiso', *Parassitologia*, *15*, 273-80

Balbo, T., Costantini, R. & Peracino, V. (1975) 'Indagine sulla Diffusione dei Nematodi Polmonari nello Stambecco (*Capra ibex*) e nel Camoscio (*Rupicapra rupicapra*) del Parco Nazionale Gran Paradiso e della Riserva di Valdieri', *Parassitologia*, *17*, 65-8

Belda, A. & Ferrando, J. (In press) 'Contribution au Diagnostique de la Kerato-conjonctivite du Chamois (*Rupicapra rupicapra*) en Espagne', *Atti Simp. Int. Cherato-congiuntivite Camoscio*, Vercelli

Biocca, E., Balbo, T. & Costantini, R. (1974) '*Nematodirus davtiani alpinus* subsp. n., Gastro-intestinal Nematode from Steinbock, *Capra ibex* and Chamois, *Rupicapra rupicapra* of the Parco Nazionale del Gran Paradiso, Italian Western Alps', *Parassitologia*, *16*, 57-62

Biocca, E., Balbo, T. & Costantini, R. (1976) '*Protostrongylus rufescens boevi* subsp. nova, Parassita dell'Apparato Respiratorio del Camoscio', *Parassitologia*, *18*, 33-40

Biocca, E., Balbo, T. & Costantini, R. (In press) 'Su due Specie nuove del Genere *Nematodirus* Parassiti di Stambecchi e Camosci: *Nematodirus ibicis* n.sp. e *Nematodirus rupicaprae* n.sp.' *Parassitologia*, *23*

Biocca, E., Balbo, T., Guarda, F. & Costantini, R. (1975) 'L'importanza della Volpe nella Trasmissione della Sarcosporidiosi dello Stambecco (*Capra ibex*) nel Parco Nazionale del Gran Paradiso', *Parassitologia*, *17*, 17-24

Cornaglia, E., Guarda, F., Guarda, F. & Misciatelli, M. (1980) 'Ricerca, Frequenza, Diagnosi, Morfologia Istopatologica ed Ultrastrutturale della Sarcosporidiosi Cardiaca nei Camosci e Stambecchi', *Ann. Fac. Med. Vet. di Torino*, *27*, 279-96

Guarda, F. & Calliera, G. (1961) 'Contributo allo Studio Anatomo-istopatologico delle Alterazioni Renali nei Camosci e Stambecchi', *Ann. Fac. Med. Vet. di Torino*, *11*, 199-213

Guarda, F., Guarda, F. & Biolatti, B. (1980) 'Contributo allo Studio dell'arteriosclerosi Aortica e Coronarica nei Camosci e Stambecchi delle Alpi', *Ann. Fac. Med. Vet. di Torino*, *27*, 303-17

Guarda, F., Guarda, F. & Cornaglia, E. (1980) 'Contributo allo Studio della Patologia Cardiaca dei Camosci e Stambecchi', *Ann. Fac. Med. Vet. di Torino*, *27*, 253-74

Maglione, E., Ginanni, C. & Sacco, T. (1969) 'Su Alcune Caratteristiche Biologiche di un Ceppo di Virus dell'Ectima Contagioso Isolato dagli Stambecchi del Parco Nazionale del Gran Paradiso', *Atti XV Cong. Naz. di Microbiologia*, Torino-Saint Vincent, pp. 1-11

Perco, F. & Tosi, G. (in press) 'Gegenwaertige Lage des Gamswildes in Italien', *Acta CIC Tagung*, Ljubljana

Tournut, J. (in press) 'Observations et Recherches sur la Kerato-conjonctivite des Isards', *Atti Simp. Int. Cherato-congiuntivite Camoscio*, Vercelli

Valenza, F., Julini, M., Cornaglia, E. & Beccaria, E. (1976) 'Sulla Broncopolmonite Necrotica dei Camosci', *Ann. Fac. Med. Vet. di Torino*, *27*, 70-85

CHAPTER 31

Do Parasites Play an Important Role in Competition between Ibex and Chamois?

W. SCHRÖDER and H. KOFLER
Institute of Wildlife Research, University of Munich, Amalienstrasse 52, D-8000 München 40 (FRG)

A rapidly increasing ibex population in Eastern Austria has apparently caused a decline of chamois.

Under such unusual circumstances where ibex and chamois show a high degree of overlap in resource use in winter, severe competition for food has been suggested as the main cause underlying the chamois decrease (Schröder & Kofler, in press), but parasites may also be considered as a possible factor in their decline.

Ibex, chamois and the study area are described in Kofler & Schröder (1985).

Gastrointestinal tracts of 22 chamois and 27 ibex were examined. The animals were of different sex and age classes, harvested during a hunting season.

Of 21 parasite species all except two were found in chamois as well as in ibex. *Nematodirus europaeus* and *Nematodirus spathiger* were found only in chamois. One species, *Nematodirus abnormalis*, was found for the first time in Austria.

The parasite loads are shown in Figure 31.1. There was no animal without any parasite. *Trichostrongylus axei* was found most often in both species (chamois 82 per cent, ibex 85 per cent) but the percentage of infection differed greatly for *Haemonchus contortus* (chamois: 63 per cent; ibex: 26 per cent), *Trichostrongylus capricola* (chamois: 41 per cent; ibex: 78 per cent) and *Chabertia ovina* (chamois: 36 per cent; ibex: 81 per cent). The

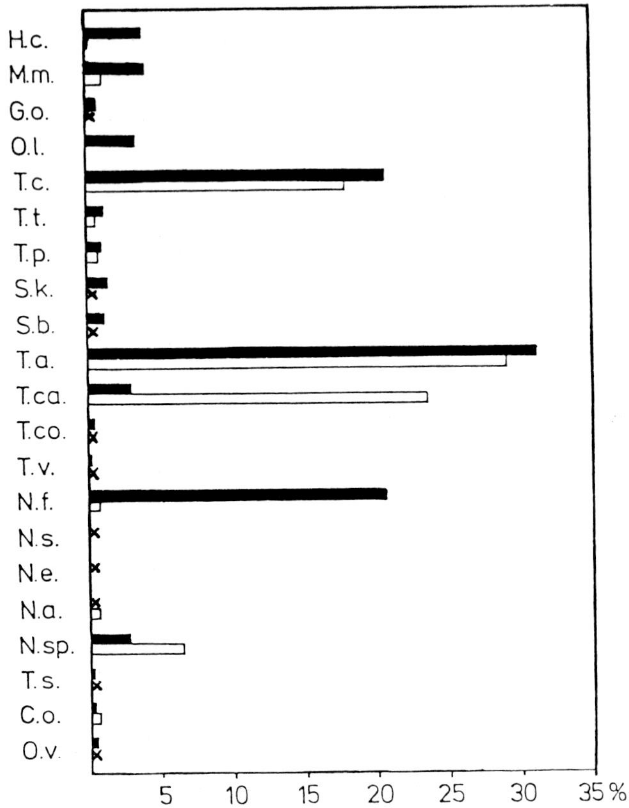

Figure 31.1: Percentage of Hosts Infected: black = chamois, white = ibex. The species names can be seen in Table 31.1

size of the parasite populations in the two hosts was estimated after Samuel *et al.* (1977) (Table 31.1). The parasite population load proved to be greater in ibex.

Ecologists and wildlife biologists often ignore the possible role of parasites in regulating host populations. Barbehenn (1969) argued that the outcome of competition between two host species can be influenced by a parasite which is more destructive to one species than to the other. If parasites play a role in the decline of chamois, this may only happen by affecting their mortality rate and/or reproduction. No evidence of a parasite-caused mortality could be found in this area and any influence on reproduction is difficult to assess.

Table 31.1: Populations of Parasites in Chamois and Ibex

Parasite species	Numbers* Chamois	Ibex
Haemonchus contortus (H.c.)	949	388
Marshallagia marshalli (M.m.)	3,180	1,779
Grossispiculagia occidentalis (G.o.)	452	296
Ostertagia leptospicularis (O.l.)	1,182	407
Teladorsagia circumcincta (T.c.)	5,456	7,381
Teladorsagia trifurcata (T.t.)	750	555
Teladorsagia pinnata (T.p.)	455	740
Skrjabinagia kolchia (S.k.)	1,227	228
Spiculopteragia böhmi (S.b.)	409	37
Trichostrongylus axei (T.a.)	6,135	8,520
Trichostrongylus capricola (T.ca.)	409	6,224
Trichostrongylus colubriformis (T.co.)	318	259
Trichostrongylus vitrinus (T.v.)	454	222
Nematodirus filicollis (N.f.)	5,722	3,408
Nematodirus spathiger (N.s.)	90	—
Nematodirus europaea (N.e.)	180	—
Nematodirus abnormalis (N.a.)	90	2,645
Nematodirus spezies (N.sp.)	1,362	5,112
Trichuris spp. (T.s.)	154	318
Chabertia ovina (C.o.)	54	326
Oesophastomum venulosum (O.v.)	136	51
Total	29,164	38,896

* = The size of the population is estimated as percentage of infected individuals × median of parasites.

Yet, chamois were observed to move out of their former range, away from ibex. Therefore, it has been assumed that heavy food competition may have occurred (Kofler, 1981). We think that such a cause is likely to be a more important factor than parasites alone in determining the chamois decline.

References

Barbehenn, K.R. (1969) 'Host-parasite Relationships and Species Diversity in Mammals: an Hypothesis', *Biotropica*, *1*, 29-35

Kofler, H. (1981) 'Okologisch-vegetationskundliche Untersuchungen zur Nahrungswahl und Konkurrenz von Gams und Steinbock im Hochlantschstock/Steiemark'. Diss. Universität Graz

Kofler, H. & Schröder, W. (1985) 'Harvesting an Atypical Ibex (*Capra ibex*) Population: A Management Plan' in Lovari, S. (ed.) *The Biology and Management of Mountain Ungulates*. Croom-Helm, London, pp. 212-15

Schröder, W. & Kofler, H. (in press) 'Coexistence and Competitive Exclusion of Chamois and Ibex', *Acta Zool. Fenn.*

Samuel, W.M., Hall, W.K., Stelfox, J.G. & Wiskart, W.D. (1977) 'Parasites of Mountain Goat of West Central Alberta with a Comparison of the Helminths of Mountain Goat and Rocky Mountain Bighorn Sheep', *Proc. 1st Int. Mountain Goat Symposium*, Kalispell (Montana), pp. 212-25

Index